MACHINE LEARNING AND IOT FOR INTELLIGENT SYSTEMS AND SMART APPLICATIONS

Computational Intelligence in Engineering Problem Solving

Series Editor: Nilanjan Dey

Computational Intelligence (CI) can be framed as a heterogeneous domain that harmonized and coordinated several technologies, such asprobabilistic reasoning, artificial life, multi-agent systems, neuro-computing, fuzzy systems, and evolutionary algorithms. Integrating several isciplines, such as Machine Learning (ML), Artificial Intelligence (AI), Decision Support Systems (DSS), and Database Management Systems (DBMS) increases the CI power and impact in several engineering applications. This book series provides a well-standing forum to discuss the characteristics of CI systems in engineering. It emphasizes on the development of CI techniques and their role as well as the state-of-the- art solutions in different real world engineering applications. The book series is proposed for researchers, academics, scientists, engineers and professionals who are nvolved in the new techniques of CI. CI techniques including artificial fuzzy logic and neural networks are presented for biomedical image processing, power systems, and reactor applications.

Applied Machine Learning for Smart Data Analysis
Nilanjan Dey, Sanjeev Wagh, Parikshit N. Mahalle, Mohd. Shafi Pathan

IoT Security Paradigms and Applications
Research and Practices
Sudhir Kumar Sharma, Bharat Bhushan, Narayan C. Debnath

Applied Intelligent Decision Making in Machine Learning
Himansu Das, Jitendra Kumar Rout, Suresh Chandra Moharana, Nilanjan Dey

Machine Learning and IoT for Intelligent Systems and Smart Applications
Madhumathy P, M Vinoth Kumar and R. Umamaheswari

Industrial Power Systems: Evolutionary Aspects
Amitava Sil and Saikat Maity

For more information about this series, please visit: https://www.crcpress.com/Computational-Intelligence-in-Engineering-Problem-Solving/book-series/CIEPS

MACHINE LEARNING AND IOT FOR INTELLIGENT SYSTEMS AND SMART APPLICATIONS

Edited by
Madhumathy P., M. Vinoth Kumar, and
R. Umamaheswari

CRC Press is an imprint of the
Taylor & Francis Group, an **informa** business

First edition published 2022
by CRC Press
6000 Broken Sound Parkway NW, Suite 300, Boca Raton, FL 33487-2742

and by CRC Press
2 Park Square, Milton Park, Abingdon, Oxon, OX14 4RN

CRC Press is an imprint of Taylor & Francis Group, LLC

Library of Congress Cataloging-in-Publication Data
A catalog record has been requested for this book

ISBN: 978-1-032-04723-2 (hbk)
ISBN: 978-1-032-04725-6 (pbk)
ISBN: 978-1-003-19441-5 (ebk)

DOI: 10.1201/9781003194415

Typeset in Times
by MPS Limited, Dehradun

Contents

Preface

This book includes all associated topics to Machine learning, Big data, Internet of things and applications in Internet of things. The main objective of this book is to fetch numerous innovative studies in Machine learning, Big data and Internet of Things. It supports the researchers, engineers and students in several interdisciplinary domains to support realistic applications. The book presents an overview of the different algorithms by focusing on the advantages, disadvantages and applications of each algorithm in the field of Machine learning and IOT. The book provides machine learning (ML) techniques to address both intelligence and configurability to various IoT devices. The book also reports the challenges and the future directions in the IoT and machine learning. This book comes with an energy-efficient cross layer model and energy-related routing metric combination to prolong the lifetime of low power IoT networks. This book deals with Machine Learning which is subset of AI that uses computational statistics to find a mathematical model describing Input and Output Data. Machine Learning techniques have been successfully involved in a various applications including assistance in medical diagnosis and analyzing disease based on clinical and laboratory symptoms with appropriate data to give more efficient result for diagnosing disease.

Though these new skills are prodigious, they result in numerous challenges including resource constraints of IoT devices, poor interoperability, heterogeneity of IoT system and several privacy and security vulnerabilities. They also expose severe IoT security challenges. Further, traditional security approaches against the most prominent attacks are insufficient. Therefore, enabling the IoT devices to learn and adapt to various threats dynamically and addressing them proactively need immediate attention. In this regard, machine learning (ML) techniques are employed to address both intelligence and reconfigurability to various IoT devices.

Outcome:

- Apply different AI techniques including machine learning and deep learning
- Perform supervised and unsupervised machine learning for IoT data
- Implement distributed processing of IoT data in AI platforms
- Implementing AI from case studies in Personal IoT and Industrial IoT

Editors' Biographies

Dr. Madhumathy P. is working as a professor at Dayananda Sagar Academy of Technology and Management, Bengaluru, Karnataka, India. She completed her engineering from Anna University in 2006. M.E (gold medalist) from AVIT in 2009 and Ph.D. from Anna University in 2015. With rich experience in teaching for about 14 years, her area of interests include Computer Networks, Wireless Communication, Wireless sensor Networks, Internet of Things, Wireless Channel Modeling, Mobile Communication and topics related to Networks and Wireless Communication domains. Having published more than 75 papers in international, national journals and conferences, she is a life member in ISTE and senior member from IEEE, and is serving as a reviewer for IEEE, IET, Springer, Inderscience and Elsevier journals. She has registered and published three Indian patent. Having received a grant for the title "A Complex Programmable Logic Device Based Green House Monitoring System for Agriculture" from VGST, Govt. of Karnataka, under SMYSR program, she has published a book titled "Computer Communication Networks" with ISBN number 978-81-937245-1-4. She has acted as publication chair for international IEEE conference held at DSATM, and has conducted and coordinated many workshops and FDPs.

Dr. M. Vinoth Kumar, obtained his Bachelor's degree in Computer Science and Engineering from Periyar University, Salem, Tamilnadu, India. He obtained his Master's degree in Computer Science and Engineering and his PhD in Computer Science majoring in Agent Programming from Anna University, Chennai, Tamilnadu, India. Currently, he is an associate professor at the Faculty of Information Science and Engineering, Dayananda Sagar Academy of Technology and Management, Bangalore, Karnataka, India. His specializations include Artificial Intelligence, Machine learning and Big Data Computing. His current research interests are convolutional neural network and medical image processing. Having published 45 research papers in reputed national, International journals and conferences, he has filed 6 innovative patents and 1 patent is granted by the Indian patent office. Reviewer and editorial member in Indexed national and International Journals, he is the life member of Computer Society of India(CSI), Indian Science Congress Association(ISCA) and associate member of Institute of Engineers(India), Indian Society of Technical Education(ISTE).

Dr. R. Umamaheswari, currently working as assistant professor in Department of Electronics & Instrumentation Engineering at SRM Valliammai Engineering College, Kattankulathur, Tamilnadu, India, has completed her Ph.D. in the field of wireless communication in the year 2017 from Anna University. She completed her Masters in VLSI Design Engineering (2011) from Anna University with a gold medal, and her Bachelors in Electronics and Instrumentation Engineering (2004) from Madras University. She has more than 10 years of teaching experience and specializes in the core area of soft computing techniques. An innovative person with deep knowledge of Artificial Intelligence, Neuro-fuzzy systems and IoT, she has published more than 25 research articles in national and international journals and published three text books for Basic electrical, electronics and instrumentation engineering for second semester Anna University syllabus. She has filed three patents in India; organized guest lecture, seminars, faculty development program under the banner of All India Council of Technical Education (AICTE); delivered guest lecture in various institutions and also shared various chair-positions in conferences and seminars. She is Life Member of professional societies like ISTE, ISC, CSI, IAENG.

Contributors

Dr. T.R. Ganesh Babu
Professor
Department of Electronics and Communication Engineering
Muthayammal Engineering College
Rasipuram, Namakkal, Tamil Nadu

Ishita Banerjee
Research Scholar
Dayananda Sagar Academy of Technology and Management

Dr. B. Chidhambararajan
Professor/Principal
SRM Valliammai Engineering College
Kattankulathur, Chennai, Tamil Nadu

K. Elaiyaraja
Assistant Professor
Department of Information Technology
SRM Valliammai Engineering College
Kattankulathur, Chennai, Tamil Nadu

Dr. R. Elakkiya
Assistant Professor
Center for Information Super Highway (CISH)
School of Computing, SASTRA Deemed to be University

Gadee Gowwrii
M.Sc. (Statistics)
Osmania University

Anuj Kumar Gupta
Professor
Department of CSE
Chandigarh Group of Colleges
Landran, Mohali, India

Asif Hasan
Assistant Professor
Department of Psychology
Aligarh Muslim University
Aligarh, India

Bilal Khan
Department of Computer Science
University of Bradford
United Kingdom

Rahul Kakkar
Associate Professor
Department of Applied Science
Chandigarh Group of College
Landran Mohali

Dr. J. Kirubakaran
Associate Professor
Department of Electronics and Communication Engineering
Muthayammal Engineering College
Rasipuram, Tamil Nadu

Dr. T. Kumanan
Principal
Faculty of Engineering and Technology
Meenakshi Academy of Higher Education and Research,
West K.K Nagar, Chennai, Tamil Nadu India

Dr. M. Senthil Kumar
Associate Professor
Department of Computer Science and Engineering
SRM Valliammai Engineering College
Kattankulathur, Chennai

Dr. Mohan Kumar S.
Professor
Nagarjuna College of Engineering and Technology
Bangalore, Karnataka, India

Dr. Tribhuwan Kumar
Assistant Professor of English
College of Science and Humanities at Sulail, Prince Sattam Bin Abdulaziz University

Dr. Vinoth Kumar M.
Associate Professor,
Department of Information Science and Engineering
Dayananda Sagar Academy of Technology & Management
Bangalore, India

Dr. P. Madhumathy
Professor, Department of Electronics and Communication Engineering
Dayanand Sagar Academy of Technology and Management
Bangalore, India

M.G.M. Milani
Faculty of Integrated Technologies
Universiti Brunei Darussalam
Bandar Seri Begawan, Brunei Darussalam

Dr. G. Padmapriya
Associate Professor
Department of Computer Science and Engineering
Saveetha School of Engineering, SIMATS
Chennai, India

Moumita Pal
Assistant Professor
Department of Electronics and Communication Engineering
JIS College of Engineering

Binay Kumar Pandey
Assistant Professor
Department of IT
College of Technology, Govind Ballabh Pant University of Agriculture and Technology
United Kingdom

Digvijay Pandey
Department of Technical Education
IET, Dr. A.P.J. Abdul Kalam Technical University
Lucknow, Uttar Pradesh, India

Dr. Balachandra Pattanaik
Professor
Department of Electrical and Computer Engineering
College of Engineering and Technology, Wollega University
Ethiopia, Africa

S. Poovizhi
Research Scholar
Department of Electronics and Communication
Engineering
Anna University
Chennai

Dr. M. Prakash
Associate Professor
Department of Computer Science and Engineering
SRM Institute of Science and Technology
Kattankulathur, Chengalpattu District, Tamil Nadu

Dr. R. Praveena
Associate Professor
Department of Electronics and Communication Engineering
Muthayammal Engineering College

Dr. R. Raja
Assistant Professor
Department of Electrical and Electronics Engineering
Muthayammal Engineering College

Murugaiya Ramashini
Department of Computer Science and Informatics, Faculty of Applied Sciences
Uva Wellassa University
Badulla, Sri Lanka

Ranjana Ray
Assistant Professor
Department of Electronics and Communication Engineering
JIS College of Engineering
Kalyani

S. Sandhya
Research Scholar
Department of Information Technology
SRM Valliammai Engineering College

Bikramjit Sharma
Assistant Professor
Department of ME
Thapar Institute of Engineering and Technology
Patiala, India

Manvinder Sharma
Assistant Professor
Department of ECE
Chandigarh Group of Colleges
India

Joginder Singh
Assistant Professor
College of Engineering and Management, Punjabi University
Patiala

V. Sudha
Assistant Professor
Department of Electronics and Communication Engineering
Sona College of Technology

Dr. Thilakavathi B.
Professor
Department of Electronics and Communication Engineering
Rajalakshmi Engineering College
Chennai, Tamil Nadu, India

Dr. Umamaheshwari R.
Assistant Professor
Department of Electronics and Instrumentation Engineering
SRM Valliammai Engineering College

Dr. Malik Mohamed Umar
School of Engineering and Applied Sciences, Kampala International University
Uganda

Randy Joy M. Ventayen
Dean
CBPA, Pangasinan University
Philippines

Dr. Vidhya K.
Professor
Department of Electronics and Communication Engineering
Saveetha School of Engineering, SIMATS
Chennai, Tamil Nadu, India

1 A Study on Feature Extraction and Classification Techniques for Melanoma Detection

S. Poovizhi[1], *Dr. T. R. Ganesh Babu*[2], *Dr. R. Praveena*[3], *and Dr. J. Kirubakaran*[4]

[1]Research Scholar, Department of Electronics and Communication Engineering, Anna University
[2]Professor, Department of Electronics and Communication Engineering, Muthayammal Engineering College
[3]Associate Professor, Department of Electronics and Communication Engineering, Muthayammal Engineering College
[4]Associate Professor, Department of Electronics and Communication Engineering, Muthayammal Engineering College

CONTENTS

DOI: 10.1201/9781003194415-1

1.1 INTRODUCTION

The incidence of malignant melanoma is in the majority cases fatal and increasing worldwide. According to the 2020 Melanoma Skin Cancer Report of the Global Cancer Observatory, there were 287,723 cases of melanoma and 1,042,056 of non-melanoma cancers recorded globally with a greater number of cases in Australia and the United States of America than anywhere else in the world [1]. Gender-wise, men are 10% more likely to develop melanoma skin cancer than women and 4% more likely to die from melanoma than women. (Figure 1.1) gives a snapshot of growth of skin cancer from current to future projected levelsand (Table 1.1) gives projected levels of new cases of skin cancer.

The two major types of skin cancers are Melanoma and Non-Melanoma. Melanoma arises from malignant melanocytic cells of the epidermis and this cell produce melanin pigment which decides the color of the skin. The malignant melanocytic cells grow abnormally and invade other skin cells forming a big mass of cells called the tumor. Early detection and treatment profoundly lead to prognosis of the disease. With many skin imaging techniques developed in recent years for assisting dermatologists to detect the melanoma, some of the techniques are (i) Total Body Photograph [2] (ii) Ultrasonography [3] (iii) Epiluminescence Microscopy [4] (iv) Cross Polarization Epiluminescence and (v) Optical Coherence Tomography. Dermatologists use noninvasive methods [5] like dermoscopic and macroscopic methods to identify melanoma in dermoscopic images. Dermoscopy provides better diagnosis as compared to that with the naked eye [6,7].

Based on the deep residual network, the Sultana et al. [8]-developed skin cancer classification system by using the regularized fisher platform and the convolutional neural network to extract the low discriminative function to classify the melanoma. Sonia et al. [9] proposed the non-subsampled contourlet transform to extract features using local binary pattern and gray level co-occurrence matrix, but due to high dimensionality, it cannot extract texture features effectively. Many approaches have been developed to identify melanoma from normal images. Zaquot et al. [10] used ABCD techniques to identify melanoma performing median filtering in pre-processing and feature extraction by entropy and bifold methods. Ma et al. [11] developed skin cancer classification system by extracting ABCDE features that contain both shape and color features. Machine learning algorithms like decision tree, K-nearest neighbors, and artificial neural networks were used by [12] to compare the performance of each algorithm for melanoma classification. Also, melanoma is identified by its seven characteristics using the so-called seven-point checklist [13] and its three characteristics using the so-called the three-point checklist [14].

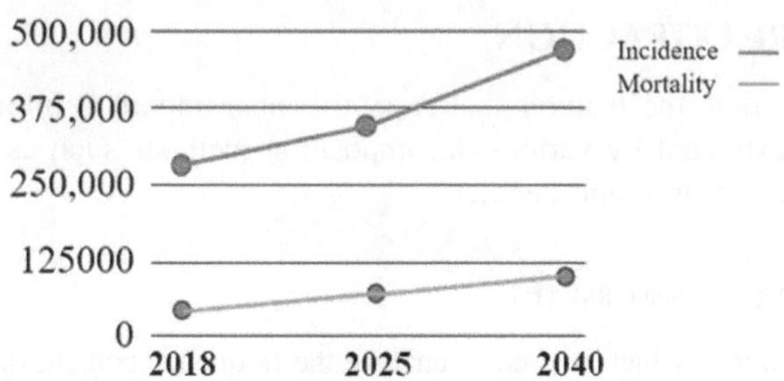

FIGURE 1.1 Incidence and Mortality Rate of Melanoma Cancer.

TABLE 1.1
Global Cancer Observatory Report 2020

Estimated New Cases in 2025	340,271
Estimated New Cases in 2040	466,914

Nasir et al. [15], who extracted three features namely color, texture and histogram-oriented gradients features for melanoma detection, selected the superior features as input to SVM classifier for the classification of melanoma images. Rahul et al. [16], who discussed a skin cancer classification system in which the preprocessing step is carried out by median filtering and enhancement using histogram equalization methods, used the residual convolutional neural network as classifier to classify the images; but with the drawback that the noise gets enhanced due to enhancement in the preprocessing stage. Wavelet based skin cancer classification system developed by [17] decomposes the image to give approximate and detailed coefficients which are ploughed back into the probabilistic neural network for further classification. Another Wavelet based resnet model was used by [18] but it requires large amounts of training data to perform better classification than expectation.

Donoho et al. [19] developed continuous curvelet transform for detecting curves in the medical image but the construction of the waveform was not built-in to work with a discrete domain directly. While deep neural network-based classification was presented by [20] for the classification of melanoma and non-melanoma images, the drawback of this method is that only biopsy proven images were considered in classification module. Deep learning methods were presented by Goyal et al. [21] for lesion segmentation in skin cancers, where the lesion are segmented exactly from the dermoscopic images and various metrics such as specificity, sensitivity and accuracy calculated to measure the performance of the method.

1.2 FEATURE EXTRACTION

In feature extraction, the features that carry the most important information about the image are extracted by various decomposition methods such as wavelet, curvelet, contourlet, shearlet and bendlet.

1.2.1 Fourier Transform (FT)

The fourier transform, which is used to analyze the frequency content of the stationary signals, decomposes the signal f(t) into the sum of its sine and cosine components.

The continuous fourier transform is given by

$$\mathrm{F\{f(t)\}} = \int_{-\infty}^{\infty} f(t)e^{-j\omega t}\mathrm{dt} \tag{1.1}$$

The inverse fourier transform is given by

$$\mathrm{F^{-1}\{f(t)\}} = \int_{-\infty}^{\infty} f(t)e^{j\omega t}\mathrm{dt} \tag{1.2}$$

The discrete fourier transform of the two-dimensional signal is given by

$$\mathrm{F(k,\ l)} = \sum_{i=0}^{M-1} \sum_{j=0}^{N-1} f(i,j)e^{-j2\pi\left(\frac{ki}{M}+\frac{jl}{N}\right)} \tag{1.3}$$

The inverse fourier transform of two-dimensional signal is given by

$$\mathrm{f(i,\ j)} = \frac{1}{\mathrm{MN}} \sum_{i=0}^{M-1} \sum_{j=0}^{N-1} F(K,l)e^{j2\pi\left(\frac{ki}{M}+\frac{jl}{N}\right)} \tag{1.4}$$

Where M, N represent variable size and i, j represent continuous variable.

Drawbacks

The Fourier Transform does not provide time information when the signal occurs but gives the frequency information that exists in the signal.

It is not suitable for non-stationary signals.

1.2.2 Short Time Fourier Transform (STFT)

In STFT, the non-stationary signals are analyzed by windowing the signal, i.e the signal is divided into segments and each segment is stationary.

STFT is represented by

$$\mathrm{STFT(t',\ f)} = \int x(t) * w(t - t'')e^{-j\omega t}dt \tag{1.5}$$

where, f(t) is the signal,
w(t) is the window function.

STFT gives poor time resolution and good frequency resolution for wide window, and poor frequency resolution and good time resolution for narrow window.

Drawbacks

- Once window size is fixed, it cannot be changed.
- STFT cannot extract texture features from images.

1.2.3 Wavelet Transform

Mallat [22] proposed a mathematical tool called wavelet transform for image processing applications where wavelets give both time and frequency information of the signal. A wavelet provides multiresolution analysis i.e represents the image on more than one scale. The advantage of multiresolution analysis is that the features undetected at one scale can be detected by the other scale. Wavelets can represent discontinuity of the image with fewer coefficients.

A wavelet is a small wave with limited duration that has zero average value and is given by

$$\Psi_{a,b}(t) = \frac{1}{\sqrt{a}}\psi\left(\frac{t-b}{a}\right)\text{a, b}\epsilon\text{R} \tag{1.6}$$

Here,

a. Scale (Dilation)
b. Translation (Position)

Generally, wavelets are characterized at scale and location. Wavelets are broadly classified as

- Continuous wavelet transform.
- Discrete wavelet transform.

The 1-D Continuous wavelet transform is given by

$$\text{CWT(a, b)} = <\text{f}, \Psi\text{a}, \ \text{b}> = \frac{1}{\sqrt{\text{a}}}\int_{-\infty}^{\infty} \psi*\left(\frac{t-b}{a}\right)dt \tag{1.7}$$

1.2.3.1 Discrete Wavelet Transform

The Discrete wavelet transform is obtained by sampling of the continuous wavelet transform. It decomposes the image into sub-bands by processing rows and columns separately and by down sampling the image by a factor of 2 using FIR filters. As a result, four bands namely LL, LH, HL and HH are obtained. LL (approximate image) contains low horizontal and low vertical frequency information that has been obtained from low pass image. The other three bands LH, HL and HH give detailed information of horizontal, vertical and diagonal directions obtained from the high pass image. (Figure 1.2) depicts discrete wavelet transform that decomposes the image into 2 sub-bands for rows and down-samples each of these sub-bands by a factor of 2 to give four sub-bands when processing columns.

Discrete wavelet transform of 2D image of size M×N is expressed as

Approximate Coefficients

$$W\varphi(j_0, m, n)\frac{1}{\sqrt{mn}}\sum_{x=0}^{m-1}\sum_{y=0}^{n-1} f(x, y)\varphi j_0, m, n(x, y) \tag{1.8}$$

Detailed Coefficients

$$W_\psi^i(j_0, m, n) = \frac{1}{\sqrt{mn}}\sum_{x=0}^{m-1}\sum_{y=0}^{n-1} f(x, y)\psi j_0^i, m, n(x, y) \tag{1.9}$$

Where, i= {H, V, D}

Drawbacks

- Wavelet transform can capture frequency content only in limited directions, so it has poor directionality.
- Inefficient for capturing anisotropic features like lines and curves.

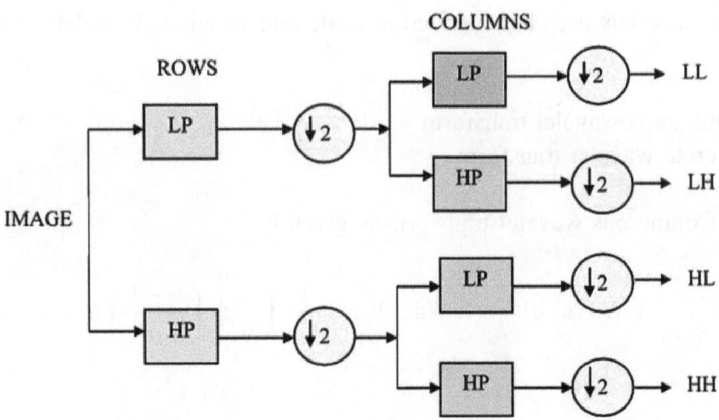

FIGURE 1.2 Wavelet Decomposition Filter Bank.

- Many coefficients are needed to capture point singularities.

1.2.3.2 Discrete Curvelet Transform

Curvelet transform [23] is a multiscale geometric transform that can represent edges and curve singularities effectively compared to wavelets. Curvelets can represent images in various levels and directions. Generally, curvelets exist at all scales, locations, and orientations of curves in the image. Curvelets very well capture anisotropic features and obey parabolic scaling law. The parabolic scaling law states

$$\text{Width} = \text{Length}^2$$

The curvelet transform is given by

$$C(j,\ \theta,\ k_1,\ k_2) = \sum_{\substack{0\le x<M \\ 0\le y<N}} f(x,\ y),\ \varnothing_{j,\theta,k_1k_2(x,y)} \tag{1.10}$$

Where, θ = Orientation

k_1, k_2 = Spatial Location of curvelets

f(x,y) = Input image having dimension M×N

Image decomposition in curvelet transform consists of following steps. In the first step, the image is decomposed into sub bands by applying wavelet transform. Then each sub band is smoothly windowed as squares by parabolic scaling and each square is applied with discrete ridgelet transform. The wedges of parabolic shape in the fourier plane are obtained by a radial window function and an angular window function. The radial window is responsible for image decomposition into scales, and the angular window is responsible for orientation and direction. Finally, curvelet coefficients are obtained by applying IFFT to all the wedges. (Figure 1.3) depicts curvelet transform as a combination of parabolic scaling of a sub-band into a smooth square window, and discrete ridgelet transform applied on each square thereafter.

Drawbacks

- Construction is easy in continuous domain and difficult to implement in discrete domain.
- Difficult to sample on the rectangular grid.
- Curvelet involves rotations and these operators will not preserve digital lattice.

1.2.3.3 Discrete Contourlet Transform

Do M N et al. [24] proposed a multiscale and multi-direction transform which is useful in capturing contours in medical images. The above depiction allows approximate images to be gained from coarse to fine resolutions. Discrete curvelet transform is constructed by combining Laplacian Pyramid (LP) and Directional Filter Bank (DFB) (Figure 1.4).

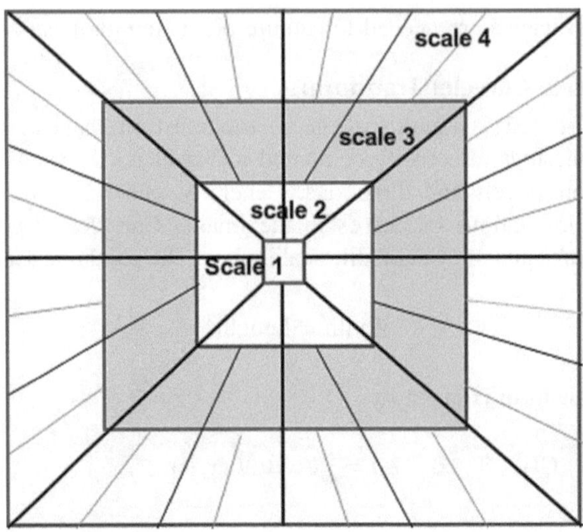

FIGURE 1.3 Frequency Tiling.

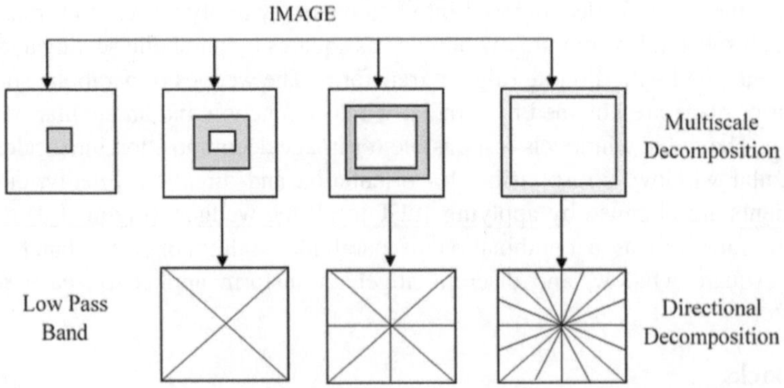

FIGURE 1.4 Contourlet Decomposition.

Laplacian Pyramid captures point discontinuities, and directional filter bank connects these point discontinuities and transforms them into linear structure. Given the original image is decomposed by laplacian pyramids into various bands of frequencies, and where the use of pyramid structure enhances the edge features in the image, the low pass band is further sub-sampled to get frequency bands. The directional filter bank is used to generate necessary directions required from each scale. Contourlets capture contours along geometric smoothness in all possible directions, so that contourlet transform performs better in capturing the intrinsic geometric features such as edges, curves and contours [25].

Drawbacks

- Contourlet transform suffers from Pseudo Gibbs effect.
- It cannot detect the non-smooth corner points.
- It is not a shift invariant.

1.2.3.4 Discrete Shearlet Transform

Shearlets which is based on multilevel and multi-direction transform are developed efficiently to analyze the anisotropic features. The important property of shearlet is that it allows optimally sparse approximation for cartoon-like functions. Shearlets are constructed by parabolic scaling, shearing, and translation. (Figure 1.5) depicts discrete shearlet transform where a shearlet is constructed at first decomposition of the image and then localized with direction information to get an optimally sparse approximation for the function.

Shearlet Transform can be written as

$$ST(\psi) = T_t D_{Aa} D_{Ss} \psi \tag{1.11}$$

Where,

T_t is the translation operator,
D_{Aa} is the dilation, and
D_{Ss} is the shearing operator.
The dilation is given by scaling matrix and the shearing matrix

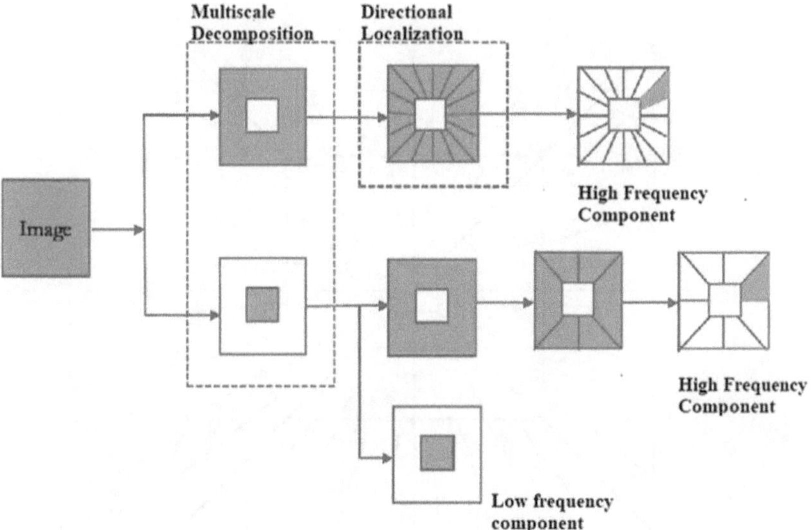

FIGURE 1.5 Shearlet Decomposition.

$$A_a = \begin{pmatrix} a & 0 \\ 0 & a^{\frac{1}{2}} \end{pmatrix} wherea > 0 \quad (1.12)$$

$$S_s = \begin{pmatrix} 1 & s \\ 0 & 1 \end{pmatrix} \text{where } s \text{ is an integer} \quad (1.13)$$

Shearlet transform uses anisotropic dilation and orientation to precisely capture the geometric edges. The continuous shearlet transform is represented as

$$\psi_{a,s,t}(x) = a^{\frac{-3}{4}} \psi(A^{-1}B^{-1}(x - t)) \quad (1.14)$$

$$SH_{\psi}f\,(a, s, b) = <f, \psi_{a,s,t}> ,a > 0, s \quad (1.15)$$

Figure 1.6 shows the frequency tiling containing horizontal part of the cone and vertical part of the cone. The directional components are obtained by the translating the shear matrix.

Drawbacks

- Shearlets cannot classify the curvature precisely.
- Shearlets are redundant in nature.

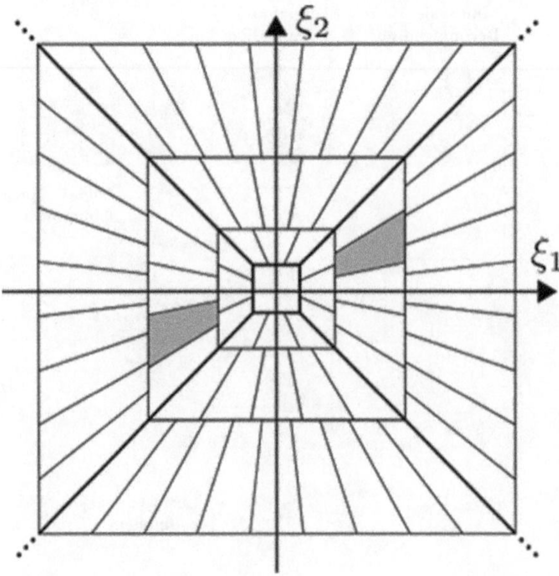

FIGURE 1.6 Frequency Tiling.

1.2.3.5 Bendlet Transform

Bendlet is a second-order shearlet transform. Bendlets can capture anisotropic features by scaling, translation, shearing, and bending parameters. The second-order shearlets classify the image more precisely compared to what the first-order shearlets do. Bendlet transform supports alpha scaling instead of parabolic scaling. The alpha scaling matrix is given by (Figure 1.7).

$$A_{a,\alpha} = \begin{pmatrix} a & 0 \\ 0 & a^{\alpha} \end{pmatrix} \text{ where } a > 0 \text{ and } \alpha \in [0, 1] \tag{1.16}$$

When

$\alpha = 1$ ➜ Isotropic Scaling
$\alpha = 0.5$ ➜ Parabolic Scaling
$\alpha = 0$ ➜ Pure Directional Scaling

1.3 CLASSIFICATION

Classification is an especially important phase in medical image analysis which predicts or classifies the unlabeled data to a set of known labeled data. Classification generally contains two phases: namely, the training and the testing phase. In the training phase, training datasets are used to build up a model whereas in the testing phase, validation of model takes place to qualify performance. Some important classifiers such as logistic regression, decision trees, K-nearest neighbors, and

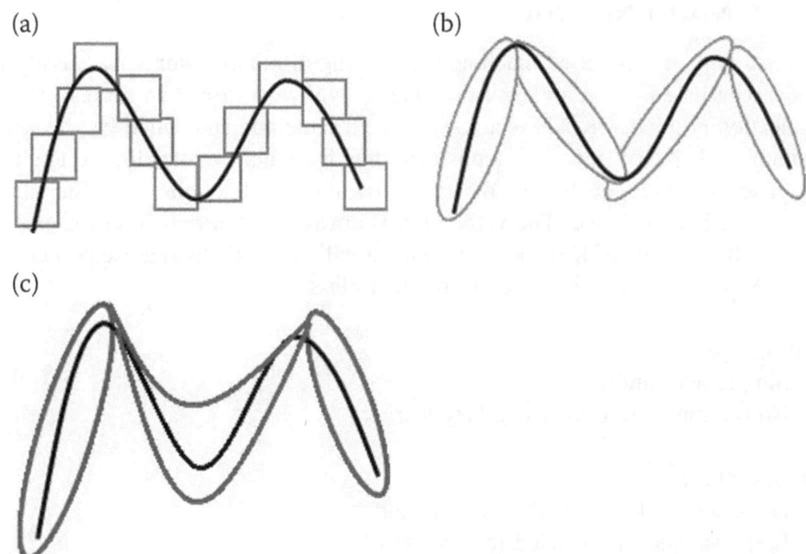

FIGURE 1.7 Curve Detection by (a) Wavelet – Many Coefficients Are Needed to Capture the Curve (b) Curvelet/Shearlet – Few Coefficients Are Needed to Capture the Curve (c) Bendlet – Very Few Coefficients Are Needed to Capture the Curve.

support vector machines are discussed which would be useful to classify melanoma from dermoscopic images into normal and abnormal images.

1.3.1 Logistic Regression

Logistic regression is a popular algorithm used for modeling classification problems. It builds the model by expressing the hyperplane that separates two datasets. It estimates the relationship between the dependent and the independent variable. The resultant outcome from logistic regression will be always categorical whereas in case of linear regression the relation is always described by continuous dependent variables. The two types of logistic regression are (i) Binary Classification and (ii) Multi class classification. Logistic regression uses sigmoid curve which maps the relation between the dependent and the independent variable, and calculates the probability that the feature vector represents an object belonging to the class.

Advantages

- Easy to implement and quite easy to train.
- Provides good accuracy when dataset is linearly separable.
- Works fast in classifying unknown records.

Disadvantages

- It cannot solve non-linear problems.
- Predicts only discrete functions.

1.3.2 K-Nearest Neighbor

K-NN is a type of supervised machine learning algorithm that works by classifying the new data based on a similarity measure. K-NN works for both regression and classification problems. K-NN is a non-parametric method and called lazy learning algorithm. In K-NN, a new instance is classified by majority votes for its neighbor class. The K-NN algorithm is measured by distant functions like Euclidean, Manhattan, Minkowski etc. The value of K is always chosen to be odd for making decisions. If the value of K is too small, then it will be sensitive to noise points, and a larger K includes majority votes from other classes.

Advantages

- Simple and intuitive.
- No training period as it is a lazy learner.

Disadvantages

- Does not work with high dimensionality.
- Large samples are needed for accuracy.

1.3.3 Decision Trees

Decision trees are supervised models where we split the data by making decisions using a series of conditions to form a tree-like structure. Decision tree classifier works for both classification and prediction. A decision tree consists of three nodes namely root node, branch node and leaf node. A root node is the initial step in the decision tree that can split into maximum branches. Root node features are extracted by attribute selection measures. An attribute selection measure is repeated till a leaf node is present. The impurity measures of decision tree are given by

$$\text{Entropy} = -\sum_{i=1}^{n} p\left(\frac{i}{t}\right)\log_2 p\left(\frac{i}{t}\right) \tag{1.17}$$

$$\text{Gini} = 1 - \sum_{i=1}^{n} p\left(\frac{i}{t}\right)^2 \tag{1.18}$$

Advantages

- Nonlinear relationship parameters will not affect performance.
- Easy to interpret.
- Need of domain knowledge is not necessary.

Disadvantages

- Always restricted to one output attribute.
- Unstable classifier depends on the dataset.
- It generates a complex tree when the dataset is numeric.

1.3.4 Support Vector Machine

SVM are supervised machine learning algorithms that work for both regression and classification. SVM works on the principle of decision planes that have decision boundaries. SVM are constructed with a hyperplane that has maximum margin. (Figure 1.8) depicts an SVM hyperplane that classifies data based on maximum margin.

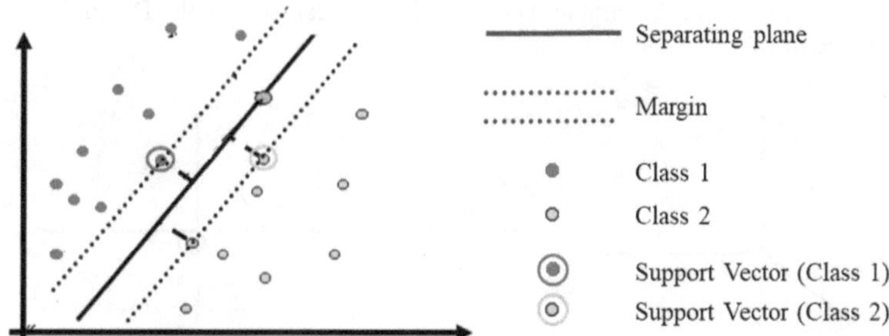

FIGURE 1.8 SVM Hyper Plane.

It consists of the separating plane that separates data into two class or multiclass. It also contains two margin lines that run parallel with hyperplane and maintains some distance to classify the data effectively. The margin lines are separated by maximum distance and each margin line passes through one of the vectors in positive and negative class. These vectors are called support vectors. Linear separable SVM can separate the data points easily by constructing a hyperplane. The data points cannot be easily separated in case of nonlinear separable SVM. (Figure 1.9) depicts SVM hyperplane for (A) linear data to be separated and (B) non-linear data.

In Non-linear case, it converts low dimension data points to high dimension datapoints using kernel function. The three types of kernel function are given below. (Figure 1.10) depicts a non-linear SVM classification using Radial Basis Function for its kernel function.

$$\text{Polynomial } \mathrm{K(a, b)} = (1 + \textstyle\sum_j a_j b_j)^d \tag{1.19}$$

$$\text{Radial Basis Functions } \mathrm{K(a,\ b)} = \exp(-(\mathrm{a} - \mathrm{b})^2/2\sigma^2 \tag{1.20}$$

$$\text{Signoid} - \text{like } \mathrm{K\ (a, b)} = \tanh(\mathrm{ca^T b} + \mathrm{h}) \tag{1.21}$$

Advantages

- Provides better accuracy.

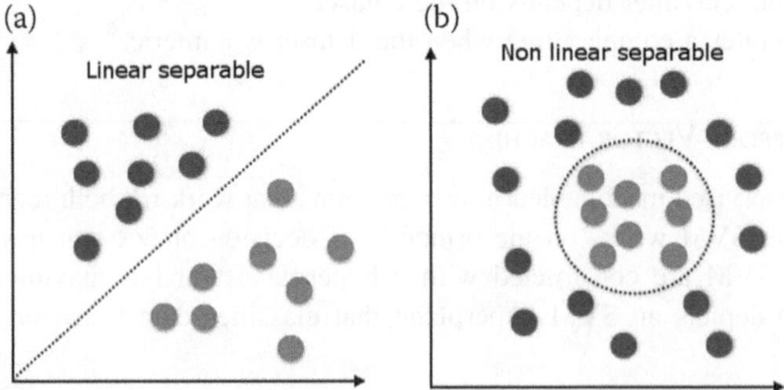

FIGURE 1.9 (a) Linear Separable Data (b) Non linear Separable Datad SP Learned Dictionaries with CS.

FIGURE 1.10 Mapping of Non-Linear Data-RBF Kernel.

- SVM works well when there is no specific trend in the datapoints.
- No overfitting problems.
- Easy to interpret nonlinear data points.

Disadvantages

- Selection of right kernels.
- Training the data takes more time.
- Computationally expensive.

1.4 SKIN CANCER DIAGNOSTIC SYSTEM FOR MELANOMA DETECTION

The aim of the proposed skin cancer classification system is to detect melanoma from dermoscopic images. Here PH2 dataset is considered for experimentation that contains 200 dermoscopic images having a resolution of 768×560 pixels. Out of 200 dermoscopic images, 120 are abnormal images and 80 are normal images. (Figure 1.11a and 1.11b) depict the classification of images into normal and abnormal.

Many research works have been proposed in recent years for the classification of normal and abnormal images. In this work, various feature extraction and classification techniques are presented, and the performance of the techniques are validated to measure the accuracy of the model.

In work I, the investigation utilizes the use of shearlet transform as feature extraction and K-Nearest Neighbors as classifier for the detection of melanoma. After finishing the preprocessing module, the features are extracted using shearlet decomposition for different scales and directions. The coefficients, which are 75 and 100 coefficients to start with, are selected from shearlet sub-bands. The selected coefficients are given as input to K-NN classifier to classify the image as normal or

(a)

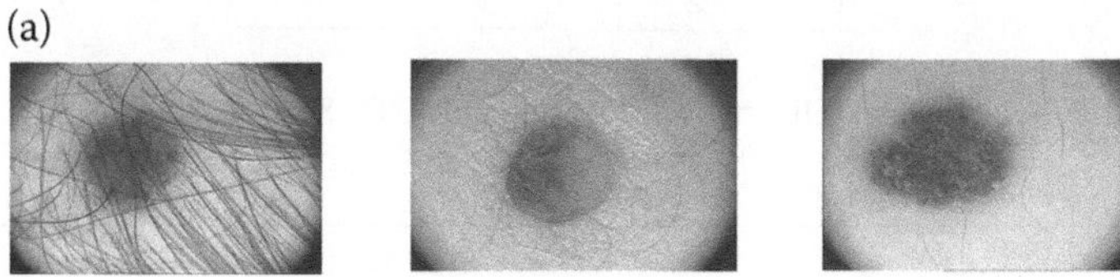

FIGURE 1.11A PH2 Datasets. Normal Images.

(b)

FIGURE 1.11B PH2 Datasets. Melanoma Images.

abnormal. (Figure 1.12) depicts the block diagram of the algorithm for classifying images in the SCC system.

The performance of the system is measured by sensitivity, specificity, and accuracy. Table 1.2 below gives the accuracy obtained from the set of 75 and 100 coefficients of shearlet transform and K-NN classifier.

It is observed from Table 1.2 that the second level of shearlet features made from 100-point coefficients yield better performance compared to other levels of shearlet decomposition. The maximum accuracy of 96% is obtained in the second level of decomposition using shearlet transform and K-NN classifier. (Figure 1.13) depicts the K-NN classifier performance for 75 coefficients vis-a-vis 100 coefficients for four levels of decomposition.

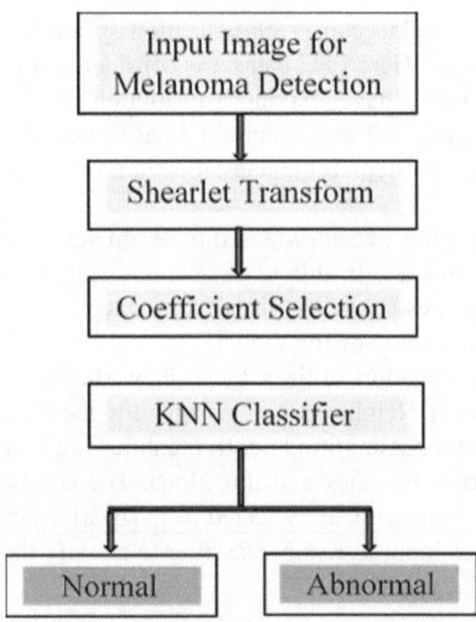

FIGURE 1.12 Block Diagram of Shearlet and KNN SCC System.

TABLE 1.2
Performance of K-NN Classifier

Shearlet Transform Decomposition Level	K-NN Classification Accuracy (%)	
	75 Coefficients	100 Coefficients
First level	82.5	88.5
Second level	88	96
Third level	81.5	91
Fourth level	85.5	92.5

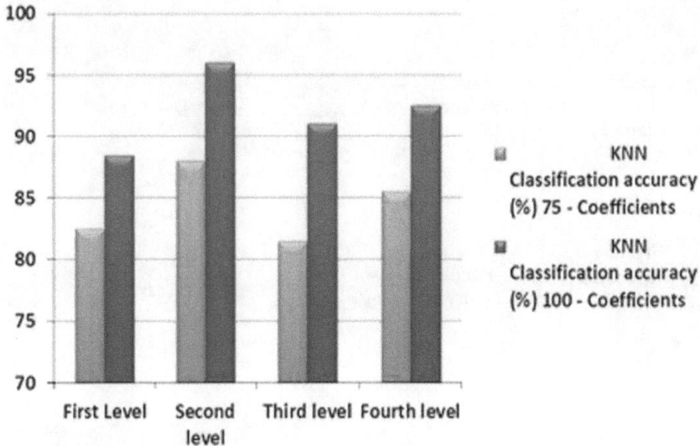

FIGURE 1.13 KNN Classifier Performance.

From the comparison of results, it is evident that the 100-coefficient extracted from the second level of decomposition produces better performance accuracy compared to the 75-coefficient from any decomposition level.

In work II, Poovizhi et al. [23] carry out bendlet transform for feature extraction and support vector machines as classifier to demonstrate how to classify normal and abnormal images from PH2 dataset. The proposed work is compared with the other image decomposition methods discussed in section 1.3. Initially the preprocessing is done by median filtering method to remove unwanted noise and hair from the dermoscopic images. The statistical features such as texture descriptors are extracted by bendlet decomposition. The energies are calculated from the coefficients of sub-bands obtained from bendlet transform. Then, the features are ploughed into support vector machines to classify normal and abnormal images. (Figure 1.14) depicts the block diagram of bendlet and SVM based SCC system.

The validation is carried out by k-fold cross validation, and the performance metrics such as sensitivity, specificity and accuracy are measured.

Where, $T_r P_o$→True Positive (#correct classification of abnormal cases),

$F_a N_e$→False Negative (#misclassification of abnormal cases),

$T_r N_e$→True Negative (#correct classification of normal cases) and

$F_a P_o$→False Positive (#misclassification of normal cases)

Table 1.3 reviews the performance of classification system for classification of normal and abnormal images. The decomposition level (from 1 to 4) and directions (2,4,8,16,32) are computed.

It is observed from Table 1.4 that bendlet transform with level 3 and direction 8 performs better with 98.50% accuracy, sensitivity of 97.50%, and specificity of 100%. (Figure 1.15) depicts the performance of various image representation techniques.

From the performance comparison of wavelet, curvelet, contourlet, shearlet and bendlet, it is bendlet transform whose sensitivity, specificity and accuracy are better compared to other representation systems.

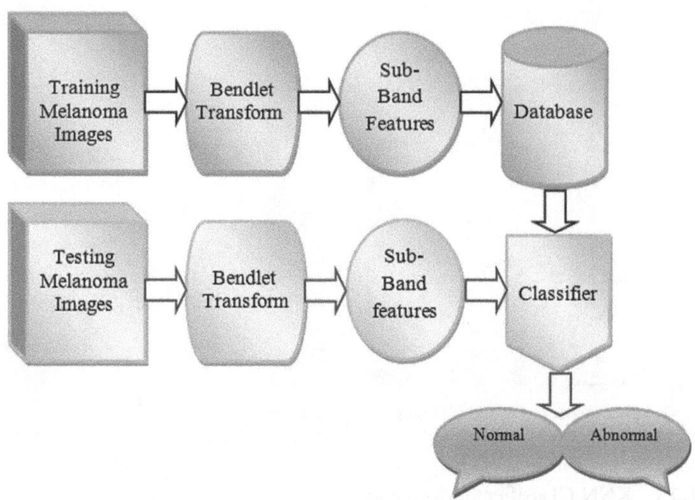

FIGURE 1.14 Block Diagram of Bendlet and SVM Based SCC System.

TABLE 1.3
Performance Metrics of SCC System

Performance Measure	Description	Formula
Sensitivity (S_n)	It is the ratio of number of true positives predicted correctly to the total number of actual positive cases.	$S_n = \frac{T_rP_o}{T_rP_o + F_aN_e}$
Specificity (S_p)	It is the ratio of the number of true negatives predicted correctly to the total number of negative cases.	$S_p = \frac{T_rN_e}{T_rN_e + F_aP_o}$
Accuracy (A_c)	It is defined as the number of skin cancer images correctly predicted to the total number of images of actual cases.	$A_c = \frac{T_rP_o + T_rN_e}{T_rP_o + F_aN_e + T_rN_e + F_aP_o}$

1.5 CONCLUSION

This study provides a brief description on various image representation systems and classification techniques used to classify dermoscopic images into normal and abnormal images. In work I, shearlet transform and K-NN classifier gives maximum accuracy of 96% in the second level of shearlet decomposition with 100 coefficients. In work II, bendlet transform for feature extraction and SVM for classification give maximum accuracy of 98.3% in level 3 decomposition and direction 8. Finally, all the image representation systems like wavelet, curvelet, contourlet, shearlet and bendlet are compared and performance metrics like specificity,

TABLE 1.4
Classifier Performance of Normal and Abnormal Phase

Level of Decomposition	Number of Directions	Performance Measures						
		T_rP_o	F_aN_e	T_rN_e	F_aP_o	S_n (%)	S_p (%)	A_c(%)
1	**2**	90	30	65	15	75.00	81.25	77.50
	4	94	26	67	13	78.33	83.75	80.50
	8	99	21	68	12	82.50	85.00	83.50
	16	94	26	68	12	78.33	85.00	81.00
	32	86	34	67	13	71.67	83.75	76.50
2	2	97	23	70	10	80.83	87.50	83.50
	4	102	18	72	8	85.00	90.00	87.00
	8	108	12	75	5	90.00	93.75	91.50
	16	100	20	71	9	83.33	88.75	85.50
	32	93	27	71	9	77.50	88.75	82.00
3	2	104	16	77	3	86.67	96.25	90.50
	4	110	10	79	1	91.67	98.75	94.50
	8	**117**	**3**	**80**	**0**	**97.50**	**100.00**	**98.50**
	16	109	11	77	3	90.83	96.25	93.00
	32	102	18	75	5	85.00	93.75	88.50
4	2	97	23	79	1	80.83	98.75	88.00
	4	108	12	79	1	90.00	98.75	93.50
	8	110	10	79	1	91.67	98.75	94.50
	16	104	16	79	1	86.67	98.75	91.50
	32	96	24	78	2	80.00	97.50	87.00

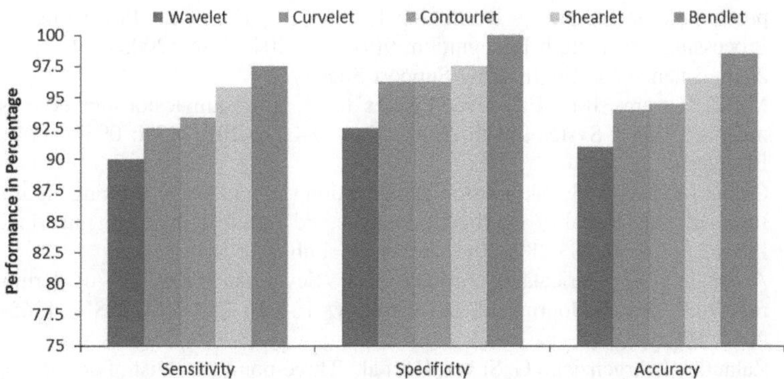

FIGURE 1.15 Performance Comparison of Various Image Representation Techniques.

sensitivity and accuracy are measured. From the results, we come to the conclusion that bendlet provides better accuracy compared to all other image representation systems. Wavelet is the worst performer whereas contourlet is the better performer than either curvelet or wavelet.

REFERENCES

[1]. del Marmol V, lipira K, '2020 Skin cancer report- global cancer observatory', 2020.

[2]. Dengel LT, Petroni GR, Judge J, Chen D, Acton ST, Schroen AT, Slingluff CL, 'Total body photography for skin cancer screening', International Journal of Dermatology, 54(11): ISSN:1365-4632, 1250–1254, 2015, Wiley.

[3]. Srivastava A, Hughes B, Hughes L, Woodcock J, 'Doppler ultrasound as an adjunct to the differential diagnosis of pigmented skin lesions', British Journal of Surgery, 73: 790–792, ISSN: 1365-2168, 1986, John Wiley & Sons.

[4]. Pehamberger H, Binder M, Steiner A, Wolff K, 'In vivo epiluminescence microscopy: improvement of early diagnosis of melanoma', Journal of Investigative Dermatology, 100(3), ISSN: 0022-202X1993, Elsevier.

[5]. Banerjee I 'Brain tumor image segmentation and classification using SVM, CLAHE AND ARKFCM', Intelligent decision support systems, applications in signal processing, 978-3-11-062110-5, 53–70, October, 2019

[6]. Banerjee I 'IOT based fluid and heartbeat monitoring for advanced health care', Classification techniques for medical image analysis and computer aided diagnosis volume 4 , ISBN: 978-0-12-818004-4, 2019 .

[7]. Madhumathy P, Prasanth R, Srilekha 'Eye movement detection for paralyzed patient using pressure sensor', International Journal of Scientific & Engineering Research, 7(11): 395–397, November, 2016 ISSN 2229-5518.

[8]. Sultana NN, Manda, B, Puhan NB, 'Deep residual network with regularised fisher framework for detection of melanoma', IET Computer Vision, 12(8): 1096–1104, ISSN 1751-8644, 2018, IET Digital Library.

[9]. Sonia R et al., 'Melanoma image classification system by NSCT features and Bayes classification', International Journal of Advances in Signal and Image Sciences, 2(2): 27–33, -ISSN: 2457-0370, 2016.

[10]. Zaqout I, 'Diagnosis of skin lesions based on dermoscopic images using image processing techniques', International Journal of Signal Processing, Image Processing and Pattern Recognition, 9(9): 189–204, ISSN: 20054254, 2207970X, 2016, Science and Engineering Support Society.

[11]. Ma Z, Tavares JM, 'Effective features to classify skin lesions in dermoscopic images', Expert Systems with Applications, 84: 92–101, ISSN: 0957-4174, 2017, Elsevier.

[12]. Ozkan IA, Koklu M, 'Skin lesion classification using machine learning algorithms', International Journal of Intelligent Systems and Applications in Engineering, 5(4), ISSN:2147-6799 285-289, 2017, Academic Publisher Science.

[13]. Argenziano G, Catricalà C, Ardigo, et al. 'Seven-point checklist of dermoscopy revisited', British Journal of Dermatology, 164(4): 785–790, ISSN: 1365-2133, 2011, Wiley.

[14]. Zalaudek I, Argenziano G, Soyer, H, et al. 'Three-point checklist of dermoscopy: an open internet study', British Journal of Dermatology, 154(3): 431–437, ISSN: 1365-2133, 2006, Wiley.

[15]. Nasir M, Attique Khan M, Sharif M, Lali IU, Saba T, Iqbal T, 'An improved strategy for skin lesion detection and classification using uniform segmentation and

feature selection based approach', Microscopy Research and Technique, 81(6):528–543, ISSN:1097-0029, 2018, Wiley.
[16]. Sarkar R, Chatterjee CC, Hazra A, 'Diagnosis of melanoma from dermoscopic images using a deep depthwise separable residual convolutional network', IET Image Processing, 13(8): 2130–2142. ISSN: 350-245X, 2019, IET Digital Library.
[17]. Jain YK, Jain M, 'Skin cancer detection and classification using Wavelet Transform and Probabilistic Neural Network', Fourth International Conference on Advances in Recent Technologies in Communication and Computing, 250-252, ISSN: 0975-8887, 2012.
[18]. SertanSerte HD, 'Wavelet-based deep learning for skin lesion classification' IET Image Processing, 14(4): 720–726, ISSN: 1350-245X, 2020. IET Digital Library.
[19]. Donoho D, Candes E, 'Continuous curvelet transform: II. Discretization and frames', Applied and Computational Harmonic Analysis, 19(2): 198–222, ISSN: 1063-5203 2005, Elsevier.
[20]. Esteva A, Kuprel B, Novoa RA, Ko J, Swetter SM, Blau HM, Thrun S. 'Dermatologist-level classification of skin cancer with deep neural networks', Nature, 542(7639): 115–118, ISSN:1476-4687, 2007.
[21]. Goyal M, Oakley A, Bansal P, Dancey D, Yap MH, 'Skin lesion segmentation in dermoscopic images with ensemble deep learning methods', IEEE Access, 8: 4171–4181, ISSN: 2169-3536, 2019.
[22]. Mallat, S, 'A wavelet tour of signal processing: the sparse way', Academic Press, United States, ISSN: 0362-4331, 2008.
[23]. Donoho D, Candes E, 'Continuous curvelet transform: II. Discretization and frames', Applied and Computational Harmonic Analysis, 19(2): 198–222, ISSN: 1096603X, 2005, Academic Press.
[24]. Do MN, Vetterli M, 'The contourlet transform: an efficient directional multi-resolution image representation', IEEE Transactions on Image Processing, 14(12): 2091–2106, ISSN: 1057-7149, 2005, IEEE.
[25]. Poovizhi S, Ganesh Babu TR, 'An efficient skin cancer diagnostic system using bendlet transform and support vector machine', Anais da Academia Brasileira de ciencias (Annals of the Brazilian Academy of Sciences), 92(1): 1–12, 2020.

2 Machine Learning Based Microstrip Antenna Design in Wireless Communications

Ranjana Ray[1], *Moumita Pal*[2], *R Umamaheswari*[3], *and Ishita Banerjee*[4]

[1]Assistant Professor, Department of Electronics and Communication Engineering, JIS College of Engineering

[2]Assistant Professor, Department of Electronics and Communication Engineering, JIS College of Engineering

[3]Assistant Professor, Department of Electronics and Instrumentation Engineering, SRM Valliammai Engineering College

[4]Research Scholar, Dayananda Sagar Academy of Technology and Management

CONTENTS

2.1 INTRODUCTION

The essential advantage in using microstrip antennas lies in the fact that they possess high power gain and omni-directional radiation patterns that are particularly useful in applications such as mobile communications, given mobile services need wideband frequency operations and will find these antennas suitable for attaining higher accuracy in their services offering. It is a critical design issue to implement printed antennas for achieving multiple application level services provided by mobile communication applications.

DOI: 10.1201/9781003194415-2

Accuracy is also a major factor to be considered in such design cases. There are multiple parameters that decide the performance of the antenna such as patch dimensions of microstrip antenna. Fabrication technique used in microstrip antenna design is easy and low cost [1,2]. For example, photo etching mechanism is another approach of fabrication of microstrip antennas which is commonly termed as patch, where various patterns and shapes for patches like square, hexagonal, rectangular, trapezoidal etc. are commonly in use. In this chapter, we have proposed an artificial neural network (ANN) modeling method for synthesis and analysis of microstrip antennas. Thus, an antenna is a radiating device that radiates electromagnetic energy in desired directions, where omni-directional; semi-directional and directional antennas are used. Antennas are used to redirect the radio frequencies provided by the transmitter to the receiver in free space [3]. Antenna is required to cover various frequency bandwidths or wide frequency band. It is expected that the antenna size should be small and its performance can be affected by changing the geometries of mounting devices [4]. Nowadays, demand for low cost antenna can be fulfilled by reduced size and compressed architecture for antennas. A microstrip antenna has several benefits when compared to supplementary antennas such as low weight, low cost, low profile, low scattering cross section, option of undeviating and rounded polarization with particular feed and can be simply incorporated for microwave circuits [5]. There are numerous procedures for investigation of microstrip antenna with some of the popular procedures being transmission line, full wave and cavity. The earliest model of microstrip antenna, among all, is the transmission line model because of its good physical insight, where the design procedure for this antenna assumed this information with dielectric constant of the substrate (ε), resonant frequency (f) and substrate height (h) [6]. A microstrip antenna is well-defined according to an array of dual radiating contracted apertures, with height h, width w and implanted L distance apart. The antenna design procedure is as follows: To find the width of the patch in practical cases for better radiation efficiencies, we follow Equation (2.1).

$$w = \frac{1}{2f\sqrt{\epsilon_o \mu_0}} \sqrt{\frac{2}{\epsilon + 1}} \tag{2.1}$$

Here $\varepsilon_o \mu_o = c = 3 \times 10^8$ m/s and f is resonant frequency.

Effective dielectric constant of antenna for w/h > 1 is given in Equation (2.2).

$$\epsilon_{reff} = \frac{\epsilon + 1}{2} + \frac{\epsilon - 1}{2} \frac{1}{\sqrt{1 + 12\frac{h}{w}}} \tag{2.2}$$

Patch width and length are calculated by Equations (2.3) and (2.4).

$$w' = \frac{v_0}{2f_r} \sqrt{\frac{2}{\epsilon_r + 1}} \tag{2.3}$$

$$l' = \frac{v_0}{2f_r\sqrt{\epsilon_{reff}}} = -2\Delta L \tag{2.4}$$

Where v_o = Velocity of the light in free space,

$$\frac{\Delta L}{h} = 0.412\frac{(\epsilon_{reff} + 0.3)\left(\frac{w'}{h} + 0.264\right)}{(\epsilon_{reff} - 0.258)\left(\frac{w'}{h} + 0.8\right)} \tag{2.5}$$

Here ΔL is the length expansion in presence fringing effects. Equation (2.6) finds the effective dielectric constant.

$$\epsilon_{reff} = \frac{\epsilon_r + 1}{2} + \frac{\epsilon_r - 1}{2}\left[1 + 12\frac{h}{w'}\right]^{-1/2} \tag{2.6}$$

Antennas have been designed for a given application according to the required performances. Nowadays, the task is to catch the available symmetrical factors of the patch (like dimensions of patch, dimensions of ground and feed position) that consume a portion of time due to trial and error practice. To decrease this consumption of time, vast amount of techniques are used (such as optimum algorithms, ANN techniques etc).

2.2 MACHINE LEARNING IN MSA DESIGN

It is necessary to develop a method to predict the resonance frequency of an antenna and artificial intelligence (AI) may be the advanced approach that is needed. AI application is found in several fields starting from communication to industrial automation with its promised reduction in costs for the users a reality. We have a detailed study of communication networks and services based on AI in [7]. Given, nowadays, these techniques in antenna design and optimization have achieved faster convergence, ML is a predictive learning based approach which is also a subset of AI which predicts outputs from its past experiences. An artificial neural network (ANN) in machine learning is an immensely comparable extended processor that has the same tendency for storing practical information and creating it accessible for usage. It looks like the human mind in two respects: first, information is attained by a net over a learning procedure, and two, neuron linking assets are identified as synaptic loads that can be used to store the observed knowledge. It has been observed that in dynamic era, artificial neural network has exceptional contributions and important advancements in the field of wireless communication. One can also request continuous target attribute regression. The results agree well with the simulation findings. Efficient design and communication call for good optimization of design parameters and researchers used ANN for optimization of return loss, antenna bandwidth, patch size, antenna gain, etc. to achieve better system models [1]. Researchers target into improving design techniques with machine learning (ML) techniques and ML is used to perform study and computation on larger empirical datasets for optimization of the ground plane slit dimensions. Also, a web repository is used to collect datasets obtained from in XML format for UWB antennas. A color display technique is also proposed, wherein firstly, a graphical tracking mechanism is followed for optimization. For USB wireless devices, the antennas used are microband ones which are good for data extraction to construct trained data by estimating fitness function using machine learning. ML can be also efficiently used for radar target classification. Antenna geometric parameter optimization as well as target classification can be performed using ML and deep learning approaches. An overview on machine learning by many researchers is given, with a major focus on investigating its usage in antenna design applications.

2.3 APPLICATION OF MSA IN IOT

Many IoT based real time applications can be modeled and implemented using ML. For example, we can discuss about weather monitoring outside the aircrafts for detecting the turbulence, pressure variations, weather outside and, accordingly, call for action. In such applications, GPS antenna is mounted on the aircraft that can range from 1.2 GHz to 1.6 GHz. Microstrip antennas can be implemented here due to its cost effectiveness, low profile and light weight. Researchers also had work based on application of Microstrip Patch Antenna in 5G communications. Currently, industries are more devoting their time on Internet of things and its application. IoT devices are most of the time needed to communicate wirelessly though it can easily access the internet. Antennas are the most indispensable part of the communication device like Smartphone, smart cities, smart watch, smart ring, smart television, smart helmet etc. where multiband antennas are useful in smart devices. Communication systems must be efficient to transfer all types of data such as voice, text and multimedia information. Currently, researchers devoting a lot of research to get improved performance from antenna have made the following progress. In [8] a new antenna design is implemented for WLAN applications with multilayer of substrate with dielectric constant. The proposed antenna shows gain 2.8 dB and return loss –20 with bandwidth of 920 MHz. Several types of antennas are in use for multiple frequency band operations such as Wi-Fi, ISM band etc. [9,10]. In [11] it is stated that for applications which require high bandwidth, a staked configuration on patch antenna can be used. The inset feeding mechanism as well as substrate materials are the key factors for improved antenna performance as shown in [12]. To support the statement we can say that 'Rogers', a substrate material, gives better performance like S11 parameters gain and return loss [13]. To filter EM noise, it's very important to design high quality transceivers that will ensure polarization and improved QoS factors of antenna [14].

IoT enables the possibility of virtual mapping between static and mobile devices. Wireless Sensor Network (WSN) is the most widely used approach for giving backbone to IoT in various fields of applications [15–17]. Antennas are omnipresent in any mode of communication, be it IoT or cellular communications. Microstrip line fed multiband microstrip patch antenna is affordable due to its low cost manufacturing and flexible design. Polarization diversity and easy to feed capacity makes it more applicable [18] as seen in applications of the antenna in fixed satellite communication, space research, to name a few. The folded miniaturized antenna for IoT (FMIoT) serves the IoT Europe band (805–835 MHz) with 95% reduced size and 55.7% radiation efficiency [19]. With reduced and folded size, a similar operation frequency range can be achieved as discussed in [20] and it promises size reduction of the physical dimensions of the antenna with slots, slits and shorting pins.

Researchers designed circular patch antenna and its characteristics specific to IoT application are analyzed for performance.

Antennas are also designed to operate in 2.4 GHz band to provide service to LTE. Also, band 12, 25 and 2.4 GHz is widely used for communications in 5G and IoT applications. The diplexer filters separate the bands according to the applications. In this chapter, design of Microstrip Antenna is optimized using ANN technique. Resonant frequency and height of dielectric substrate of Microstrip Antenna are determined by applying dimension (W,L) at the input of Artificial Neural Network optimizer. Design is initially investigated using simulation software IE3D. The results of analytical investigation are used as training data set of ANN while some are kept to be used as test data set.

2.4 DESIGN & ANALYSIS OF MSA USING ANN

2.4.1 Artificial Neural Network

Multilayer Perception Neural Networks (MLPNNs) have a vast field of applications [21,22]. It follows the back propagation algorithm. Similar to any neural network, it also consists of input layer, a hidden layer and the output layer. The inputs of hidden layer is assigned from the input layer as x_j [23]. The neurons in hidden layer aggregates x_j values and assigns weight to obtain output data y_i as shown in Equation (2.7).

$$y_i = f\left(\sum w_{ij} x_j\right) \tag{2.7}$$

Here *f notation* represents hyperbolic or sigmoid tangent function. Similarly, the output layer values are also aggregated.

ANN works on adjusting the weights of neurons using learning algorithm. Back propagation learning algorithm [24] can make the alteration in $w_{ij}(k)$ *between* i and j as shown in Equation (2.8).

$$\Delta w_{ij\,(k)} = \alpha \delta_i x_j + \mu \Delta w_{ij}(k-1) \tag{2.8}$$

Here α is the learning coefficient, x_j is input value, μ is momentum coefficient, and δ_i depends on the position of the neuron i whether it is a hidden neuron or an output one [25,26]. Gradient descent algorithm can be used to train the NN.

Artificial Intelligence in the Estimation is shown in Equations (2.9) and (2.10).

$$y_i = f\left(\sum w_{ij} x_j\right) \tag{2.9}$$

$$\|w_{ij(k)} = \alpha\delta_i x_j + \mu\|w_{ij}(k - 1) \quad (2.10)$$

In Figure 2.1 the network consists of 3 input nodes, 7 hidden nodes and 2 output nodes. Sigmoid function is used for activation and Bias input is considered as 1 for both layers.

Here the learning rate is considered in a range of 0 and 1 with random weights. Following are the steps followed in the proposed work:

Step 1:"2" *Feed forward computation*:

1. *First layer i.e. input layer output is determined.*
2. *The inputs of the intermediate layers are found.*
3. *Hidden layer outputs are aggregated.*
4. *Inputs for the third layer i.e. the output layer is found.*
5. *Final output from the output layer is obtained.*
6. *Error prediction takes place.*

Step 2:"2" *Back propagation*

Backward error calculation from output layer to input layer.
Weight updating takes place based on learning rate.

1. *Connecting weights updating is done between of the output layer and the hidden layer neurons.*

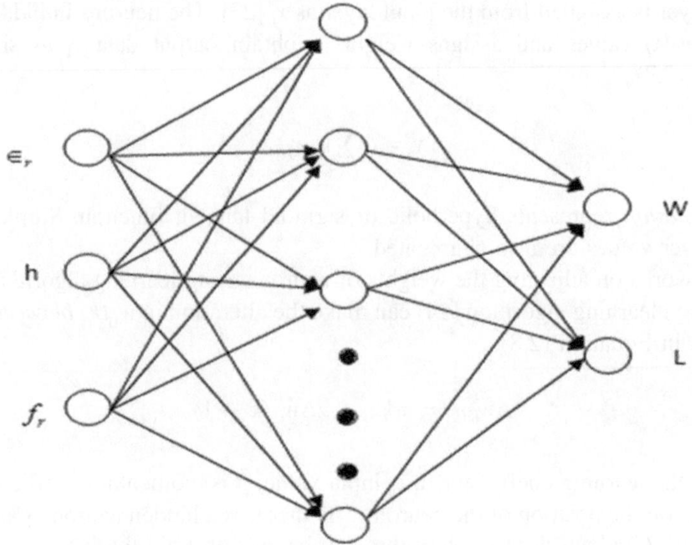

FIGURE 2.1 Example for Designing ANN Algorithm.

2. *Also update connecting weights between input layer and the hidden layer neurons.*

Step 3: *Keep repeating step 1 and step 2 to update weights at occurrence of each event* (Tables 2.1 and 2.2).

Step 4: *Terminate the process when negligibly small value of the error is obtained* (Figures 2.2 and 2.3).

2.5 RESULTS AND DISCUSSION

Optimization and simulation is performed by HFSS software for the proposed antenna model. The prototype is then used for experimental analysis. For training

TABLE 2.1
Measurement of the proposed antenna

Substrate Size	**18 mm × 20 mm**
Substrate Height	2 mm
Microstrip Feed Line length	2 mm
Microstrip Feed Line width	0.5 mm
Patch length	14 mm
`Patch width	12 mm

TABLE 2.2
Comparson of results of ANN output W,L with rest to $\in_r$, h, f_r

$\in_r$	h	f_r	W(Simulated)	L(Simulated)	W(ANN)	L(ANN)
2.6	1.9	7.5	12.1	14.5	12	14.32
2.8	1.8	6.8	12.4	14	12.49	14.23
2.1	2	6.9	11.97	14.2	11.82	14.11
2.4	1.74	7.12	12	14.3	12.2	14.9
2.93	2.1	7.3	12.7	14.4	12.56	14.32
2.7	1.9	7.5	12.63	14	12.5	13.86
2.2	2	7.9	12.5	13.9	12.72	13.9
2.51	1.8	8	12.3	13.83	12.19	13.78
3	1.6	7.7	12	13.5	11.97	13.3
2.2	1.5	7.43	12.2	14.3	12.12	14.32
2.27	1.9	7.5	12.63	14	12.5	13.86
2.82	2.1	7.3	12.7	14.3	12.12	13.77
2.1	1.989	6.9	11.97	14.2	11.82	14.11
2.024	1.5	7.43	12.2	13.378	12.42	14.12

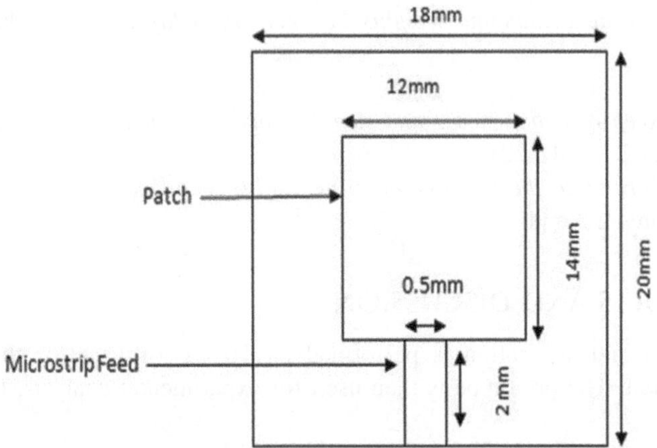

FIGURE 2.2 Geometry of Rectangular Patch Antenna.

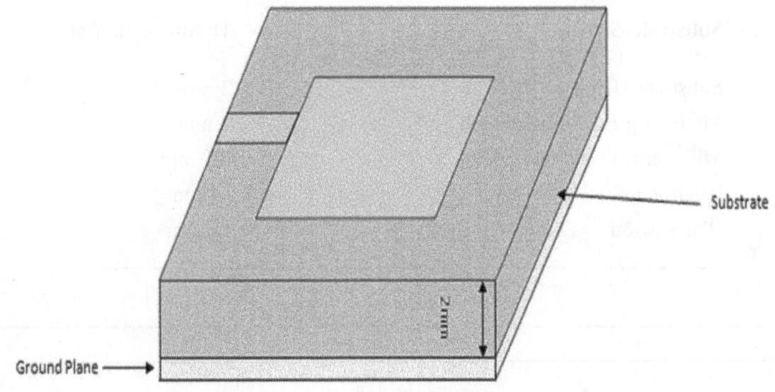

FIGURE 2.3 Front View of Rectangular Patch Antenna.

and testing the network, past data sets are used. Theoretical results are matched with measured results with maximum extent. Error obtained is infinitesimally small (Figures 2.4 and 2.5).

2.6 DESIGN OF MICROSTRIP ANTENNA AND CHARACTERIZATION USING SVM METHOD

Support vector machine (SVM) can be considered as a precise and stiff tool in antenna design. SVMs are widely in application in designing rectangular patch antenna/ rectangular patch antenna array. With proper training, SVMs are highly

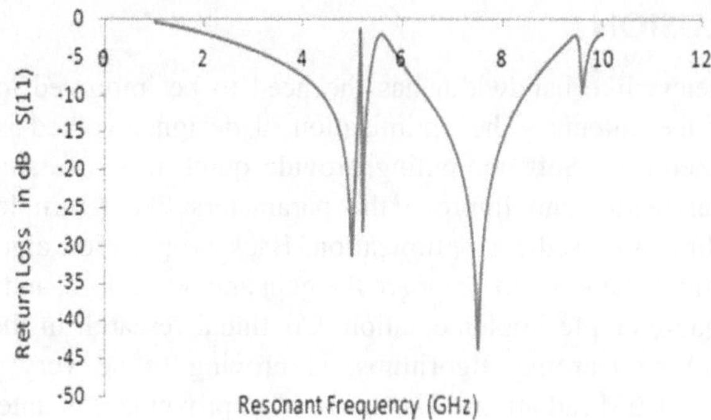

FIGURE 2.4 Return Loss vs. Resonating Frequency of Microstrip Antenna.

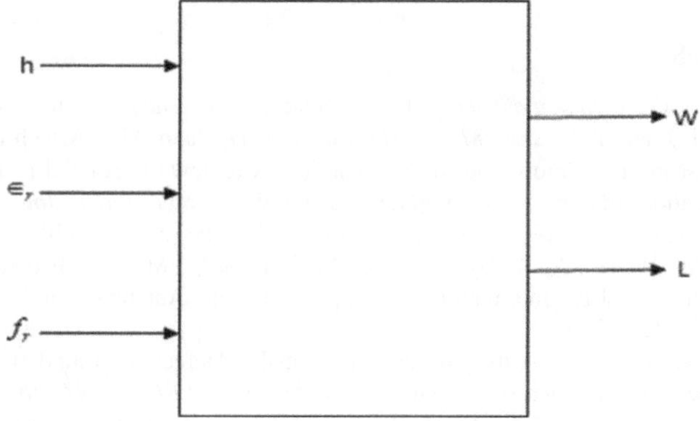

FIGURE 2.5 ANN Model.

effective in predictive antennas model design with high accuracy and precision. The structural risks are reduced and good generalization can be done using SVM.

2.7 DESIGN OF MSA FOR IOT APPLICATIONS

Patch dimensions of rectangular microstrip antennas aim to achieve the pattern maximum to be normal to the patch. It is a crucial factor to analyze microstrip antenna design factors since it operates in narrow bandwidth and also operates in close vicinity of resonant frequency. Several research works are performed on conventional rectangular slotted microstrip patch antenna and a promising precise performance is observed. The proposed work aims to provide the antenna with dual resonant frequencies of 2.39 GHz–3.15 GHz which can effectively be used IoT applications. Researchers also designed and simulated circular-shaped micro-strip patch antenna for IoT device, where performance of antenna is analyzed through S1, 1 parameter and gain.

2.8 CONCLUSION

Antenna parameters like bandwidth has the need to be improved to haul up the performance of the antenna. The optimization of design involved is a time consuming task. Therefore, Soft computing provide quick response and accuracy to optimize the bandwidth and improve the parameters like return loss, gain etc. Various algorithms are used for optimization. Back propagation algorithm can be preferred to optimize bandwidth, improve the gain and return loss, and various other parameters through simple implementation. Continual research in the field of AI, majorly in machine learning algorithms, is proving to be very promising in application areas of EM radiations. The scope of improvement in antenna design is an important area of research such as improvement in radar image classification which can even make modifications in antenna size or design area and by identifying new antenna structures.

REFERENCES

[1]. Balanis C. A. *Antenna Theory*. United States: John Wiley & Sons, Inc; 1997.

[2]. Bahl I. J. and P. Bhartia. *Microstrip Antennas*. Dedham, MA: Artech House; 1980.

[3]. Burrascano P. S. Fiori, and M. Mongiardo. "A review of artificial neural networks applications in microwave computer-aided design." *International Journal of RF and Microwave Computer-Aided Engineering*, Vol. 9, Issue 3, pp. 158–174, 1999.

[4]. Weedon W. H., W. J. Payne, and G. M. Rebeiz, MEMS switched reconfigurable antennas, IEEE International Symposium on Antennas and Propagation, pp. 654–657, 2001.

[5]. Arya A. K., M.V. Kartikeyan, and A. Patnaik. "Defective ground structure in the perspective of microstrip antenna: A review." *FREQUENZ Journal of RF-Engineering and Telecommunications*, Vol. 64, Issue 5-6, pp. 79–84, June, 2010.

[6]. Yildiz C., S. Gultekin, K. Guney, and S. Sagiroglu. "Neural models for the resonant frequency of electrically thin and thick circular microstrip antennas and the characteristic parameters of asymmetric coplanar waveguides backed with a conductor." *AEU-International Journal of Electronics and Communications*, Vol. 56, Issue 6, pp. 396–406, 2002.

[7]. Kim Y., S. Keely, J. Ghosh, and H. Ling, "Application of artificial neural networks to broadband antenna design based on a parametric frequency model." *IEEE Transactions on Antennas and Propagation*, vol. 55, pp. 669–674, Mar. 2007.

[8]. Hussein A. and S. Luhaib. "Designing E-shape microstrip patch antenna in multilayer structures for WiFi 5GHz network", 2012 20th Telecommunications Forum (TELFOR), 2012. University of Belgrade.

[9]. Ali Z., V. Singh, A. K. Singh et al., "Compact dual band microstrip patch antenna for WiMAX lower band application" In the proceedings of IEEE International Conference on Control, Computing, Communication and Materials-2013. Allahabad, Chennai.

[10]. Siddhartha M., K. Akash, and A. K. Singh, "Dual Band Textile Antennas for ISM Bands" In the proceedings of IEEE International Conference on Control, Computing, Communication and Materials-2013. Allahabad, Chennai.

[11]. Hsuan-Yu C., C. Sim, and L. Ching-Her "Compact size dual-band antenna printed on flexible substrate for WLAN operation," Antennas and Propagation (ISAP), 2012 International Symposium on, vol., no., pp. 1047,1050, Oct. 29 2012-Nov. 2 2012. Japan.

[12]. Singh V., Z. Ali, and A. Singh, "Dual wideband stacked patch antenna for WiMax and WLAN Applications", 2011 International Conference on Computational Intelligence and Communication Networks, 2011. Gwalior, India.

[13]. Sharma M., A. Katariya, and R. Meena, "E shaped patch microstrip antenna for WLAN application using probe feed and aperture feed", 2012 International Conference on Communication Systems and Network Technologies, 2012. Rajkot, Gujarat, India.

[14]. Ali Z., V. Singh, A. Singh, and S. Ayub, "E-shaped microstrip antenna on rogers substrate for WLAN applications", 2011 International Conference on Computational Intelligence and Communication Networks, 2011. Gwalior, India.

[15]. Zanella A., N. Bui, A. Castellani, L. Vangelist, and M. Zorzi. "Internet of things for smart cities." *IEEE Internet of Things Journal*, Vol. 1, Issue 1, February 2014.

[16]. Elavarasi C. and T. Shanmuganantham, "SRR loaded periwinkle flowershaped fractal antenna for multiband applications." *Microwave and Optical Technology Letters*, Vol. 59, Issue 10, 2518–2525, 2017.

[17]. Sharma P. and R. L. Yadava "Comparative analysis of feeding techniques for micro strip patch antenna and smart antenna applications in mobile communication." *IOSR Journal of Electronics and Communication Engineering (IOSR-JECE) e-ISSN: 2278-2834*, p- ISSN: 2278-8735, Vol. 9, Issue 3, Ver. IV, pp. 25–30, May-June, 2014.

[18]. Stutzman L. S. and G. A. Thiele *Antenna Theory and Design*. John Wiley, University of Dayton. IOT 8, 2002.

[19]. Regulations from Electronic Communications Committee. "The suitability of the current ECC regulatory framework for the usage of Wideband and Narrowband M2M in the frequency bands 700 MHz, 800 MHz, 900 MHz, 1800 MHz, 2.1 GHz and 2.6 GHz", *Electronic Communications Committee Report 266,* June 2017. Europe.

[20]. Wongand K. L. and K. P. Yang, "Modified planar inverted F antenna." *IEEE Electronics Letters*, Vol. 34, pp. 7–8, Jan. 1998.

[21]. Guney K., S. Sagiroglu, and M. Erler. "Generalized neural method to determine resonant frequencies of various microstrip antennas." *International Journal of RF and Microwave Computer-Aided Engineering: Co-sponsored by the Center for Advanced Manufacturing and Packaging of Microwave, Optical, and Digital Electronics (CAMPmode) at the University of Colorado at Boulder*, Vol. 12, Issue 1, pp. 131–139, 2002.

[22]. Dhaliwal B. S. and S. S. Pattnaik. "Performance comparison of bio-inspired optimization algorithms for Sierpinski gasket fractal antenna design." *Neural Computing and Applications*, Vol. 27, Issue 3, pp. 585–592, 2016.

[23]. Mishra R. K. and A. Patnaik. "Neural network-based CAD model for the design of square-patch antennas." *IEEE Transactions on Antennas and Propagation*, Vol. 46, Issue 12, pp. 1890–1891, 1998.

[24]. Devi S., Panda, D. C., and Pattnaik, S. S. (2002, June). A novel method of using artificial neural networks to calculate input impedance of circular microstrip antenna. In IEEE Antennas and Propagation Society International Symposium (IEEE Cat. No. 02CH37313) (Vol. 3, pp. 462–465). IEEE. USA.

[25]. Suma M. R. "Hybrid cloud- Intra domain data security and to address the issues of interoperability." *International Journal of Recent Technology and Engineering*, Vol 8, Issue IS 5, ISSN:2277-3878, pp. 340–344, June 2019.

[26]. Chetioui M., Boudkhil, A., Benabdallah, N., and Benahmed, N. (2018, April). Design and optimization of SIW patch antenna for Ku band applications using ANN algorithms. In 2018 4th International Conference on Optimization and Applications (ICOA) (pp. 1–4). IEEE. India.

3 LCL-T Filter Based Analysis of Two Stage Single Phase Grid Connected Module with Intelligent FANN Controllers

Dr. R. Raja[1], *Dr. V. Sudha*[2], *Balachandra Pattanaik*[3], *and Madhumathy P.*[4]

[1]Assistant Professor, Department of Electrical and Electronics Engineering, Muthayammal Engineering College

[2]Assistant Professor, Department of Electronics and Communication Engineering, Sona College of Technology

[3]Professor, Department of Electrical and Computer Engineering, College of Engineering and Technology, Wollega University

[4]Professor, Department of Electronics and Communication Engineering, Dayananda Sagar Academy of Technology and Management

CONTENTS

DOI: 10.1201/9781003194415-3

3.1 INTRODUCTION

The importance of power generation in an inverter connected to the grid tied system used direct interconnection to the grid and Renewable Energy System. SE (Sustainable Energy) or Renewable energy has recently gained attention around the world as power generation using these resources is environmentally friendly and is of less cost compared to traditional fuels. Sustainable energy sources produced high quantities of power and with better regulation, given these energy sources can be effectively utilized and can convert energy into power which can be connected to the power grid. Power electronics-based circuits are needed to execute necessary functions.

The initial function of the converter is by increasing input DC voltage from input side and its conversion into AC with constant amplitude ranges. Another function is to produce maximum power from PV source materials. Finally, AC voltage and AC current have the less Harmonic components which are minimized by using line filters. Given the AC voltage and current contains more harmonics due to switching operation, the filters reduce the harmonics. The Regulated AC output is given to the grid at a constant frequency.

The controller is used to operate the system as a closed loop function. The grid voltage is sensed and monitored by the controller to regulate the output voltage and grid voltage. The intelligent controller (Fuzzy Tuned Artificial Neural Network) is used for closed loop operation of the two-stage inverter system. The proposed inverter performance is analyzed and compared with that of the conventional controller.

3.2 LITERATURE SURVEY

Many researchers focus on other energy sources like PV (Photo Voltaic), Fuel Cells, Solar and Wind Turbines that are already popular due to the environmental issues related with the prevalent use of fossil fuels. Chen et al. [1] used high efficient grid based module along with switching technique, Zero Voltage Switching (ZVS). The stability analysis of the system with capacitor current closed loop damping technique was designed. Hybrid control in the DC-AC stage and average modeling was also presented. LLC topology was analyzed and demonstrated with MPPT technique. The performance of controller and converter was not presented.

Mahesh Kumar et al. [2] have demonstrated the DC micro-grids, where the grid connected system using non-conventional energy sources under different loads was measured. The strength of the control algorithm under various operating conditions including fault situation and efficiency was analyzed. However, fault analysis efficiency was not proved.

Claudio A Busada et al. [3] have discussed the grid connected type of filter, Inductance, Capacitance and Inductance(LCL), which has current injected and thereby, acts as a linear controller. The performance of the controller was tested on linear and nonlinear loads. It was found on comparison about performance of controllers with simulation and experimental results were not matched.

Ammar Hussein Mutlag et al. [4] designed Adaptive Fuzzy Logic Controller design method with PV system using Differential Search Algorithm. The amplitude and frequency were improved using optimized output of DSA. The design and accuracy results of the fuzzy logic controller were not accurately discussed. Remus Narcis Beres et al. [5] have developed and implemented three-phase grid tied inverter operated as Voltage Source Inverter. Switching sequence of the inverter was operated using PWM controller.

Xu Renzhong et al. [6] have presented the three phase PV based grid tied LCL filter. The design and modeled equations were used to calculate the L and LCL component values. The static and dynamic performances of the LCL filter connected inverter were not presented. Lin Chen et al. [7] have described the three phase two stage grid connected with an LC filter. The proposed technique was used to reduce the charge per watt and boost the scalability of the system. The developed ZVS technique with two stage grid connected system did not provide better consistency results.

Nascimento Filho et al. [8] have designed new methodology for current controller and DC tied voltage controller with LCL output filter. Only the dynamic behavior of the inverter was presented, because it failed to present the static results. Sandeep et al. [9] have demonstrated the design of single-phase grid tied PV based system. The inverter was operated using current feedback controller for active damping. The design procedure of the current injected in grid was presented. The current and voltage controller are used for the closed loop operation. Also, active and passive filter with LCL filter were given. Panda et al. [10] have discussed the single-phase grid connected system. PLL controller is applicable for the closed loop operation and Dual Transport Delay Based PLL (DTDPLL) technique was used. [11–14]

It was inferred that due to switching, sufficient accuracy results were also not obtained. Voltage regulation plays a significant role in single-phase inverter with LCL output filter [15–18].

3.3 PROPOSED SYSTEM

In the first stage, the PV panel provides DC power supply and it is fed to the converter. The converter boosts the PV power which is fed to the single-phase full bridge inverter. In the next stage, using inverted gate switching pulse, conversion of DC to AC voltage happens. The phase output voltage of inverter is coupled with LCL-T filter, as shown in Figure 3.1.

The switching frequency can control the output voltage as the phase difference varies between the inverters. Instantaneous power from PV panel is found by Maximum Power Point Tracking Algorithm. (Trabelsi et al. 2011). When PV panel voltage is greater than reference voltage, the DC-link regulator output voltage gets increased which results in large current flow.

Figure 3.2 shows inverter fed grid. The resonant converter circuit consists of inductance L_{S1}, capacitor C_P and series inductance L_{S2}. M_1-M_4 is switching device

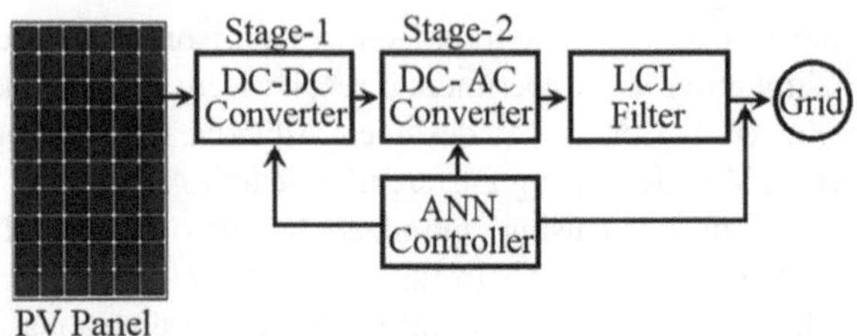

FIGURE 3.1 Block diagram of the proposed system.

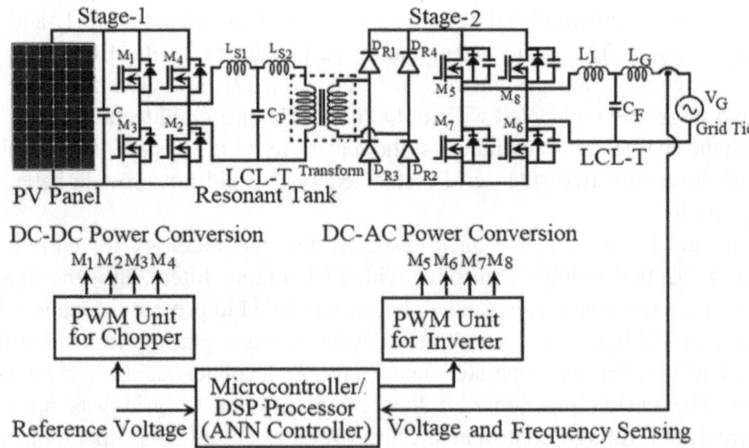

FIGURE 3.2 Two stage single phase grid connected inverter.

that operates as DC-DC converter and has gate turn off and gate turn on capabilities. The gate pulses of M_3 and M_4 are out of phase by 180° and gate pulses of M_1 and M_2 are in phase. M_5-M_8 Switches operate as inverter and output is connected to LCL-T filter.

3.3.1 Mode of Operation-1: (T_0-T_1)

At time interval (t_0-t_1), the switches M_1, M_2, M_5 and M_6 are on, and V_i PV source power is applied to inverter circuit and LCL-T resonant filter. C_p, the capacitor, gets charged; given L_{s1}, the inductor, the stage one results in current at inductor being equal to zero. At t_0, the resonant current becomes zero and capacitor, C_P is at negative value.

3.3.2 Mode of Operation-2: (T_1-T_2)

At t_1, voltage across C_P passes to positive side. At t_2, the switches M_1 and M_2 do not conduct, and across the diodes, D_1 and D_2, the inductor current flows. The diodes start to conduct. The capacitor, C_P, gets discharged. The voltage across the transformer is found positive.

3.3.3 Mode of Operation-3: (T_2-T_3)

At time interval (t_2-t_3), the switches M_3, M_4, M_7 and M_8 are turned on during which the D_1 and D_4 diodes are off, and the power is provided from the basic PV voltage source Vi to the grid. The resonant filter is enabled by negative voltage.

3.3.4 Mode of Operation-4: (T_3-T_4)

The mode of operation for stage 4, the switches M_3, M_4, M_7, and M_8 operate in non-conduction mode. The diodes D_2 and D_3 are conducting and discharge the capacitor from LS_1 side. During the interval t_4, the voltage at CP reaches zero in negative side.

3.3.5 Mode of Operation-5: (T_4-T_5)

The stage 5 process is similar to that of the stage 1 process. The diodes D_2 and D_3 are turned off as current t_5 produces change in current direction. The inverter current reaches zero and a new cycle starts.

3.4 STATE SPACE MODELING AND LCL-T FILTER DESIGN

The LCL-T filter equivalent model is shown in Figure 3.3. The inverter side inductor is L_I and the inductor L_G is across the grid, C_F is the output capacitor, V_I is given as input from the PV panel, where, V_G is voltage across the grid.

To simplify the analysis of the basic LCL-T inverter, the following assumptions are considered:

i. Ideal components are switches, diodes, inductors, and capacitors.
ii. Snubber capacitor effect is neglected.
iii. Tank circuit losses are neglected.

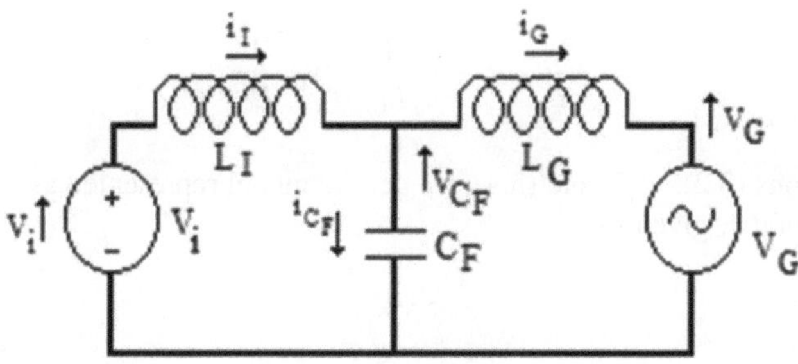

FIGURE 3.3 Equivalent circuit model of LCL–T filter.

iv. DC supply preferred.
v. Primary components of the waveforms are used for analysis.
vi. Turns ratio h = 1 for ideal transformer high.

Equivalent circuit of LCL-T Inverter is shown in Figure 3.3.

State space model is a mathematical model of state variables, where output and input are related by differential equations. The number of states, outputs, inputs, the vector variables, algebraic equations and differential equations are written as matrix equations. The time-domain approach (also known as the "state space representation") examine systems with multiple inputs and outputs. The circuit contains the number of inductors and capacitors. The state variables signify the magnetic and electric fields of the inductors and capacitors, respectively. The state equation provides system behavior. The state model gives the time history of the state variables.

$$\dot{X} = Ax + Bu \tag{3.1}$$

Where

$$\dot{X} = \begin{bmatrix} \frac{di_{L_I}}{dt} \\ \frac{dV_{C_F}}{dt} \\ \frac{di_{L_G}}{dt} \end{bmatrix} x = \begin{bmatrix} i_{L_I} \\ V_{C_F} \\ i_{L_G} \end{bmatrix} u = \begin{bmatrix} V_i \\ V_G \end{bmatrix}$$

The state- space mathematical model for LCL-T filter is found from Figure 3.3 and the load side equations are given in Equations (3.2)–(3.4)

$$\frac{d}{dt} i_{L_I} = V_i \frac{1}{L_I} - V_{C_F} \frac{1}{L_F} \tag{3.2}$$

$$\frac{d}{dt} V_{C_F} = \frac{1}{C_F} (i_{L_I} - i_{L_G}) \tag{3.3}$$

$$\frac{d}{dt} i_{L_G} = -V_G \frac{1}{L_G} + V_{C_F} \frac{1}{L_G} \tag{3.4}$$

From Equations (3.2), (3.3) and (3.4), we get the model represented as a continuous state space model, given by

$$\begin{bmatrix} \frac{di_{L_I}}{dt} \\ \frac{dV_{C_F}}{dt} \\ \frac{di_{LG}}{dt} \end{bmatrix} = \begin{bmatrix} 0 & -\frac{1}{L_I} & 0 \\ \frac{1}{C_F} & 0 & -\frac{1}{C_F} \\ 0 & \frac{1}{L_G} & 0 \end{bmatrix} \begin{bmatrix} i_{L_I}(t) \\ V_{C_F}(t) \\ i_{L_G}(t) \end{bmatrix} + \begin{bmatrix} \frac{1}{L_I} & 0 \\ 0 & 0 \\ 0 & -\frac{1}{L_G} \end{bmatrix} \begin{bmatrix} V_i \\ V_G \end{bmatrix} \tag{3.5}$$

Then consider A, B and u is written as

$$A = \begin{bmatrix} 0 & -\frac{1}{L_I} & 0 \\ \frac{1}{C_F} & 0 & -\frac{1}{C_F} \\ 0 & \frac{1}{L_G} & 0 \end{bmatrix} B = \begin{bmatrix} \frac{1}{L_I} & 0 \\ 0 & 0 \\ 0 & -\frac{1}{L_G} \end{bmatrix} u = \begin{bmatrix} V_i \\ V_G \end{bmatrix}$$

The summation between zero-input response and zero- state response for LCL-T filter is given by

$$\dot{X}(t) = \varnothing(t)A(x(s)) + L^{-1}(\varnothing(t)B(u(s))) \tag{3.6}$$

The Equations (3.5) and (3.6) can be solved in relevance to current, and the voltage components can be written as

$$\begin{bmatrix} i_{L_I} \\ V_{C_F} \\ i_{L_G} \end{bmatrix} = \begin{bmatrix} \cos\theta + \frac{1}{C_F L_G \alpha^2}[\cos\theta + 1] & -\frac{1}{L_I\alpha}\sin\theta & \frac{1}{C_F L_I \alpha^2}[\cos\theta + 1] \\ -\frac{1}{C_F\alpha}[\sin\theta] & \cos\theta & -\frac{1}{C_F\alpha}\sin\theta \\ \frac{1}{C_F L_G \alpha^2}[\cos\theta + 1] & \frac{1}{L_G\alpha}\sin\theta & \frac{1}{C_F L_I \alpha^2}[\cos\theta + 1] + \cos\theta \end{bmatrix}$$

$$\begin{bmatrix} i_{L_I}(t) \\ V_{C_F}(t) \\ i_{L_G}(t) \end{bmatrix} + \begin{bmatrix} \frac{1}{L_I}\left[\frac{1}{\alpha}\sin\theta + \frac{1}{C_F L_G \alpha^2}\left(t + \frac{1}{\alpha}\sin\theta\right) + V_i\frac{1}{C_F L_I^2}\left(t + \frac{1}{\alpha}\sin\theta\right)\right] \\ -\frac{1}{\alpha C_F}\sin\theta - V_G\frac{1}{\alpha L_G \alpha}\sin\theta \\ V_i\frac{1}{\alpha^2 L_I C_F L_G}\left[\left(1 + \frac{\sin\theta}{\alpha}\right)\right] - V_G\frac{1}{L_G}\left[\frac{\sin\theta}{\alpha} + \frac{1}{\alpha^2}\left(t + \frac{\sin\theta}{\alpha}\right)\right] \end{bmatrix} \tag{3.7}$$

Consider

$$t_1 = \frac{1}{L_I}\left[\frac{1}{\alpha}\sin\theta + \frac{1}{C_F L_G \alpha^2}\left(t + \frac{1}{\alpha}\sin\theta\right) + V_i\frac{1}{C_F L_I^2}\left(t \frac{1}{\alpha}\sin\theta\right)\right] \tag{3.8}$$

$$t_2 = -\frac{1}{\alpha C_F}\sin\theta - V_G\frac{1}{\alpha L_G \alpha}\sin\theta \tag{3.9}$$

$$t_3 = V_i \frac{1}{\alpha^2 L_I C_F L_G}\left[\left(1 + \frac{\sin\theta}{\alpha}\right)\right] - V_G \frac{1}{L_G}\left[\frac{\sin\theta}{\alpha} + \frac{1}{\alpha^2}\left(t + \frac{\sin\theta}{\alpha}\right)\right] \quad (3.10)$$

and θ_{p-1} and θ_p are the time to start and end in continuous mode conduction.

On Solving the Equations (3.5), (3.6) and (3.7) we get

$$\begin{aligned} i_{L_I}(t) = {} & i_{L_I}(\theta - \theta_{t-1})[\cos(\theta - \theta_{t-1}) + \tfrac{1}{C_F L_G \alpha^2}[1 + \cos(\theta - \theta_{t-1})]t_1 \\ & + [-\tfrac{1}{L_I \alpha}\sin\theta](V_{C_F}(\theta - \theta_{t-1}))\, t_2 \\ & + \tfrac{1}{C_F L_I \alpha^2}[1 + \cos(\theta - \theta_{t-1})] i_{L_G}(\theta - \theta_{t-1})]t_3 \end{aligned} \quad (3.11)$$

$$\begin{aligned} V_{C_F}(t) = {} & i_{L_I}(\theta - \theta_{t-1})\left[-\tfrac{1}{C_F \alpha}\sin(\theta - \theta_{t-1})t_1 + V_{C_F}(\theta - \theta_{t-1})\cos\theta\right]t_2 \\ & + i_{L_G}(\theta - \theta_{t-1})\left[-\tfrac{1}{C_F \alpha}\sin(\theta - \theta_{t-1})\right]t_3 \end{aligned} \quad (3.12)$$

$$\begin{aligned} i_{L_G}(t) = {} & i_{L_I}(\theta - \theta_{t-1})\left[\tfrac{1}{C_F L_G \alpha^2}(1 + \cos(\theta - \theta_{t-1}))\right]t_1 \\ & + V_{C_F}(\theta - \theta_{t-1})\left[\tfrac{1}{L_G \alpha}\sin(\theta - \theta_{t-1})\right]t_2 + i_{L_G}(\theta - \theta_{t-1})[\cos(\theta - \theta_{t-1}) \\ & + \tfrac{1}{C_F L_I \alpha^2}(1 + \cos(\theta - \theta_{t-1}))t_3 \end{aligned} \quad (3.13)$$

The output current, voltage and stability of the proposed system can be estimated from Equations (3.5), (3.11), (3.12) and (3.13). Based on the above equations C_F, L_1, L_G values are calculated.

3.4.1 Stability Analysis

The diagram represents LCL-T inverter from the state space Equation (3.5). The −1 + j0 is pointed in the direction of time. Consequently, if the network encirclement is valued as zero, then, the open loop has no poles in the half of s-plane. Nyquist plot for LCT-T inverter is shown in Figure 3.4.

LCL-T inverter circuit is accomplished as stable. LCL-T inverter frequency response is shown in Figure 3.5. LCL-T inverter is Stable for 35° phase margin and 17 dB gain margin.

Figures 3.6 shows the relation between converter gain (M) and frequency ratio (ω_s/ω_r) characteristics for $\delta = 1$ (L_1/L_G) Curves showing margin values.

3.4.2 Design of FANN Controller

FANN controllers can reduce steady state, dynamic and transient performance. The performance of the proposed FANN controller is improved with small neuron

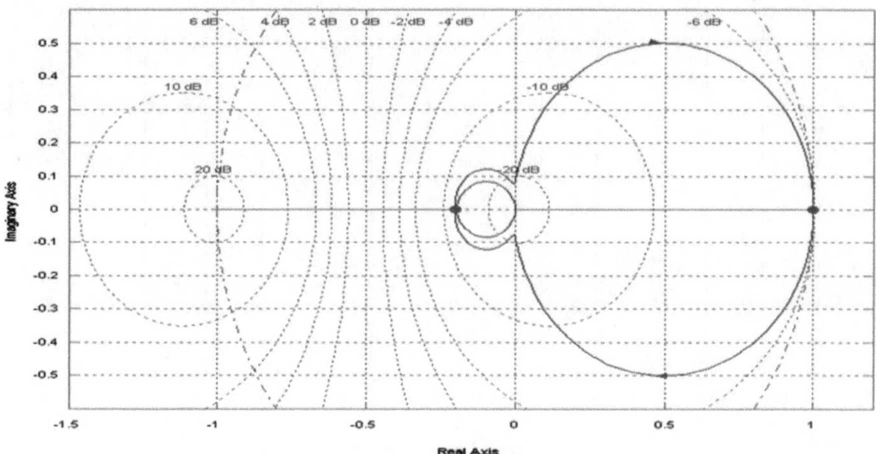

FIGURE 3.4 Stability analysis of LCL-T inverter by using Nyquist technique.

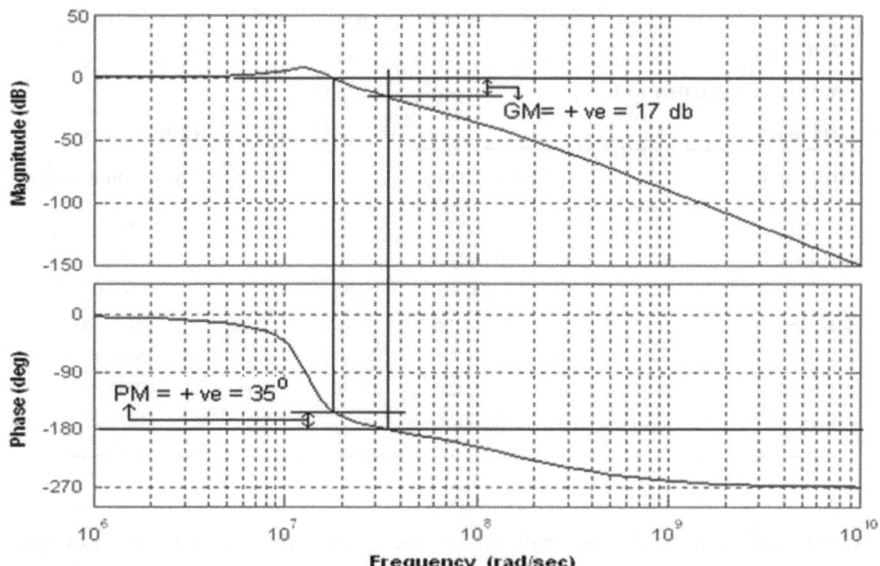

FIGURE 3.5 Frequency response of LCL-T SPRC.

structure. The drawback of fuzzy logic control scheme is the high computation level in the interpretation process.

A FLC controller was developed for two phase single stage LCL-T inverter. The proposed controller system performance was analyzed with MATLAB/Simulink for transient and dynamic execution. The Fuzzy Logic controlled inverter was simulated for 5 seconds through compile time of 0.0001 seconds. Completely 50000 data, in sequence, are obtained from the FLC. Only 6000 data are chosen from

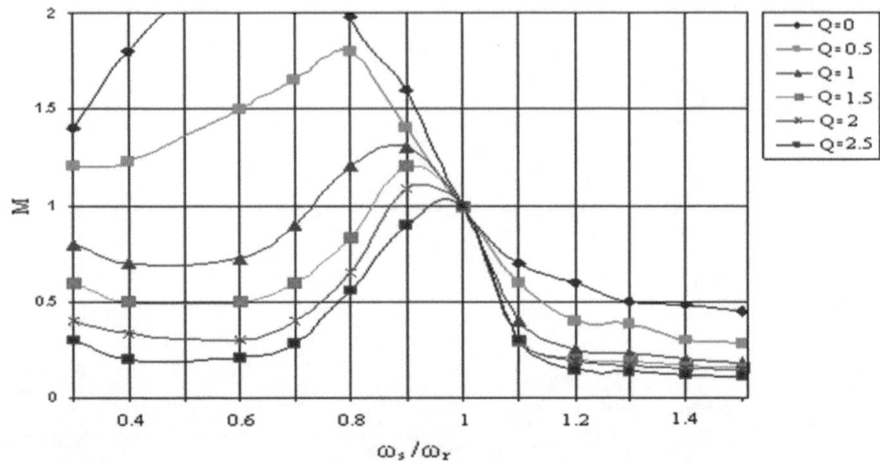

FIGURE 3.6 Control characteristics curves (M vs ω_s /ω_r).

TABLE 3.1
Sample data from FLC

Data Input		Output Data
Error Input	**Error Change**	**Corresponds to δ**
0.741823	−0.00029	−1402.75
0.659006	−0.00034	−12981
0.550268	−0.000196	−11902.7
0.479752	−0.000195	−4947.14
0.409921	−0.00021	−2068.13
0.350542	−0.00021	−4989.59
0.309802	−0.00021	−3972.25

50000 to design the FANN controller by removing the redundant data. The example data are given in Table 3.1

The proposed simple FANN controller is designed with a small number of neurons, along with a hidden layer. The input layer has two neurons formed from the feed forward NN (Neural Network), one in the output layer and the third neuron in the hidden layer.

The e(k) is the error and Δe(k) is the change in error and are the two inputs of the designed system, and the neurons are biased appropriately. The Δdc(k) is the duty cycle changes of the output network. The designed FANN is trained for error goal of 0.00596325 at 11 epochs.

The ranges for error and change in error are −1.0 to +1.0 and −2 to +2, respectively. The complete configuration of the trained network with the weights and

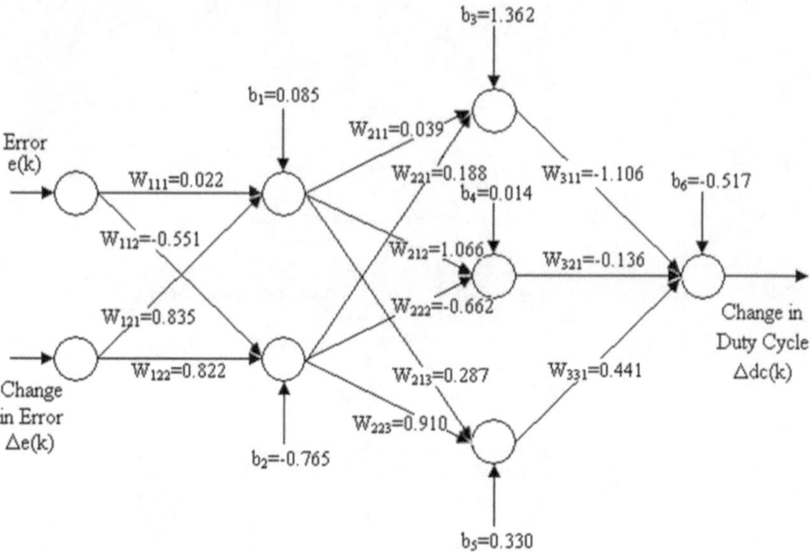

FIGURE 3.7 FANN trained network.

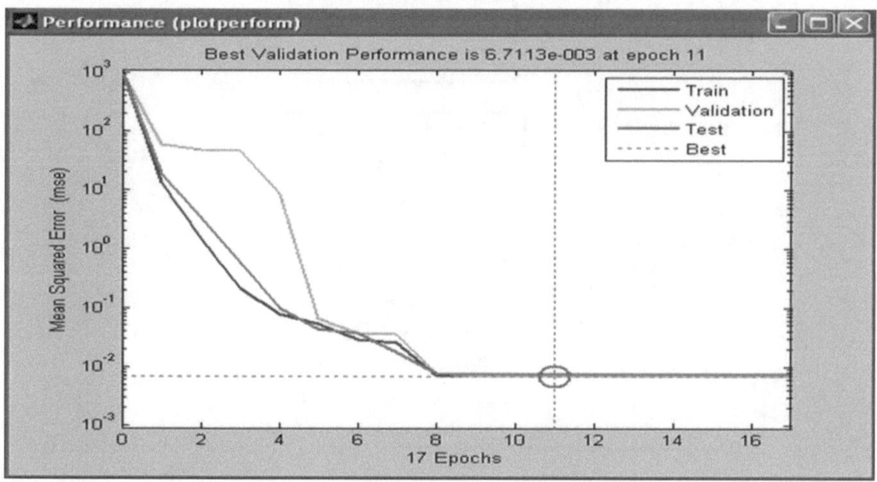

FIGURE 3.8 Performance analysis plot of FANN during training.

bias using data from the fuzzy controller are shown in Figure 3.7. (Figure 3.8 shows the performance plot of the FANN controller during training.)

The controller flow diagram works as follows (see Figure 3.9)

1. Initially from fuzzy output, the input error and change in error is found.
2. ANN is trained using back propagation algorithm.
3. The output and errors are calculated; thereby, the weights are updated.

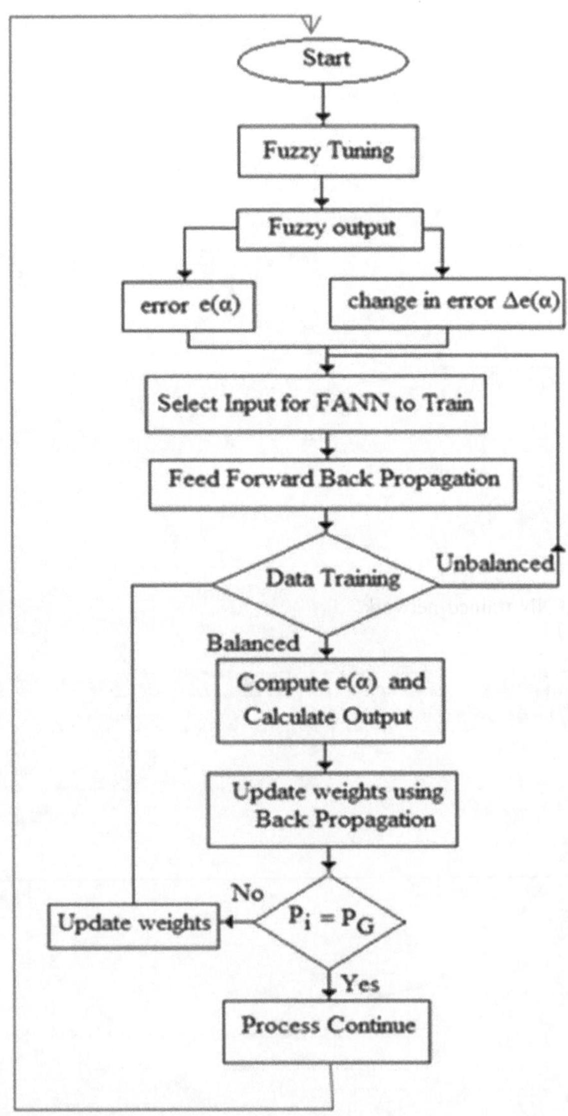

FIGURE 3.9 Flow chart of the FANN controller.

4. When training set is equal to input, the process is continued; else, wait until weights are updated.

3.5 SIMULATION RESULTS

The PV based single phase inverter grid connected system, simulated with FANN controller is carried out using MATLAB. The ANN controller function blocks are available in MATLAB, and are used to produce new inverter switching frequency for PWM pulse generators. FANN controller with inverter fed grid tied using simulation is shown in Figure 3.10.

The inverter output voltage with PID and FANN controller, respectively, with LCL-T filter response, are shown in Figures 3.11 and 3.12, where the output voltage contains harmonics and oscillation. LCL-T filter in FANN controller performed better and eliminates distortion in the output of the inverter.

The steady state error using PID was 0.4 and for proposed FANN controller it was 0.02 V. The THD value, when the LCL-T based single phase inverter is operated in PID controller, was 9.5% and for the proposed FANN controller it was 2.05%. (Figures 3.13 and 3.14, respectively, show the inverter output, the former for PID Controller; and the latter, for FANN Controller)

The Figures 3.15 and 3.16, respectively, show the transformer primary, secondary voltage and current waveforms. It seems that the overshoot and the settling time are less; and on comparing with PID controller, the response has less oscillation. Figure 3.16 shows the transformer voltage and current obtained from inverter bridge circuit.

The grid output and inverter output current response under step load conditions are shown in Figures 3.17 and 3.18 respectively. It is observed from these figures that during overload condition, the inverter current, and grid current are in the same phase. The inverter and grid output current response under load conditions are shown in Figures 3.19 and 3.20 respectively. It is observed from these figures that the grid and inverter current response are better when operated as the proposed FANN controller, given the oscillation and settling time of the current is better with less steady state error.

THD, Total harmonic distortion, of the output current in inverter is 0.39% using minimum load and 0.35% for maximum load condition. The proposed FANN controller improves the dynamic response and reduces the THD value within specified limit. The steady state error comparison for the inverter using conventional and proposed controller is given in Figures 3.21 and 3.22 respectively.

Figure 3.21 represents steady state error of PID controller with inverter output voltage.

Figure 3.22 shows the steady state error in FANN controller for inverter output voltage.

The error voltage is nearly 0.1 V at operating condition of the system.

Table 3.2 shows the rise and settling time for FANN controller compared with the conventional PID controller. Table 3.3 shows that percentage overshoot is reduced and the %THD is lower using FANN controller. (Figures 3.23 and 3.24, respectively, depict the former with (A) the experimental model with solar panel; and (B) the assembled parts and recording of outputs using oscilloscope; and the latter with driving square pulses of PWM.)

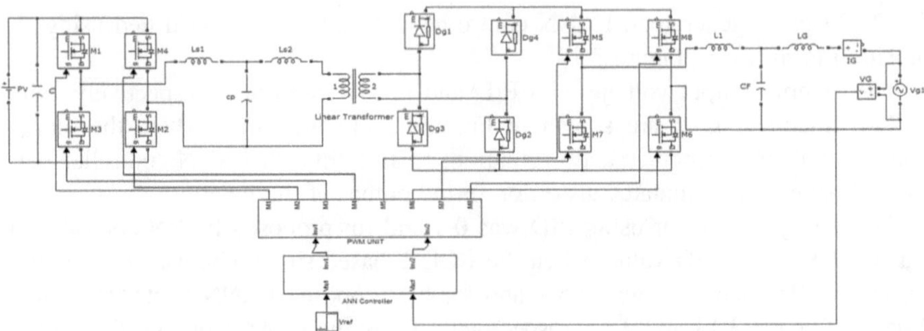

FIGURE 3.10 Simulink model of the proposed system.

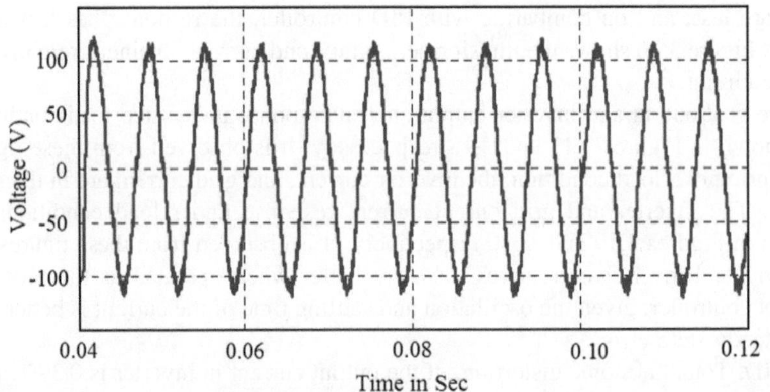

FIGURE 3.11 Inverter output (PID controller).

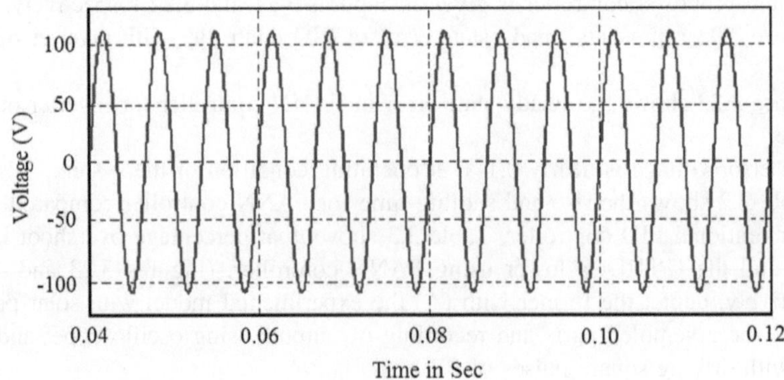

FIGURE 3.12 Inverter output (FANN controller).

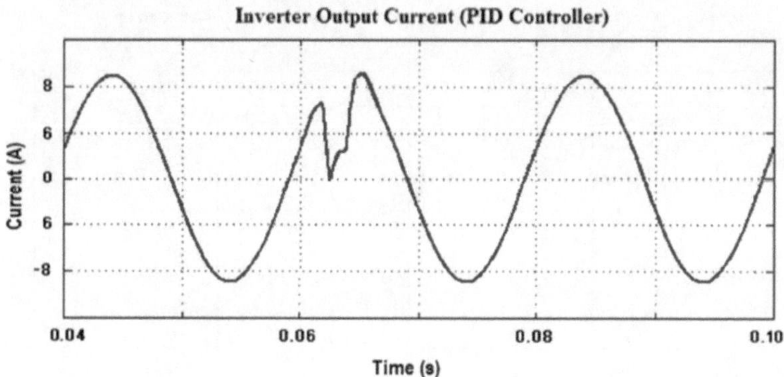

FIGURE 3.13 Output current of the inverter (PID) conventional controller, grid sisturbance at 0.06 seconds.

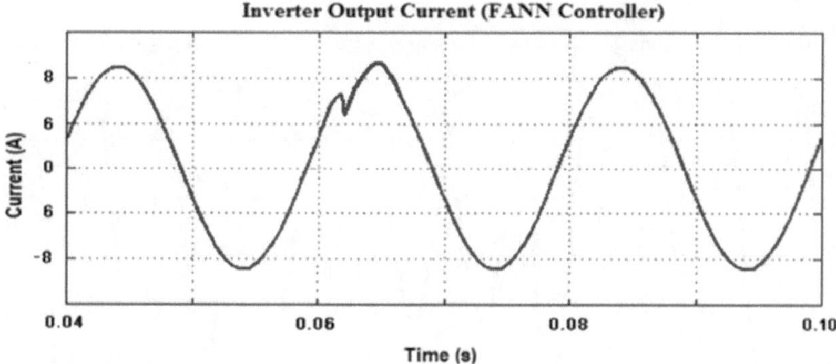

FIGURE 3.14 Output current of the inverter with FANN controller, grid disturbance at 0.06 seconds.

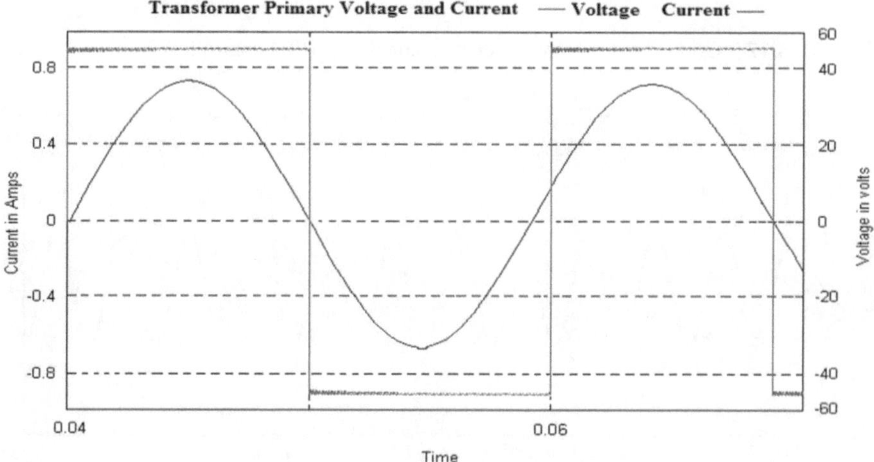

FIGURE 3.15 Transformer voltage and current primary side.

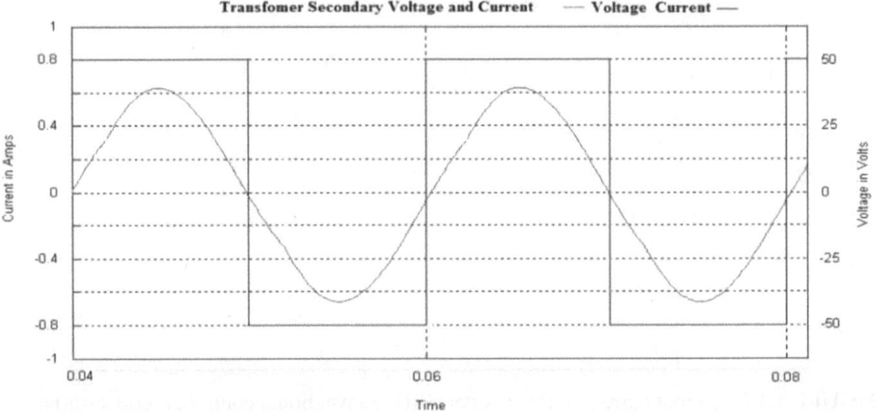

FIGURE 3.16 Transformer voltage and current secondary side.

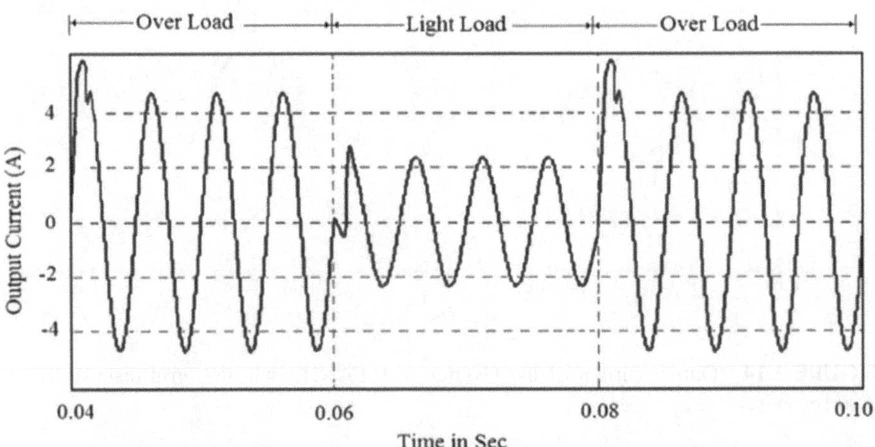

FIGURE 3.17 Grid output current under load conditions (PID controller).

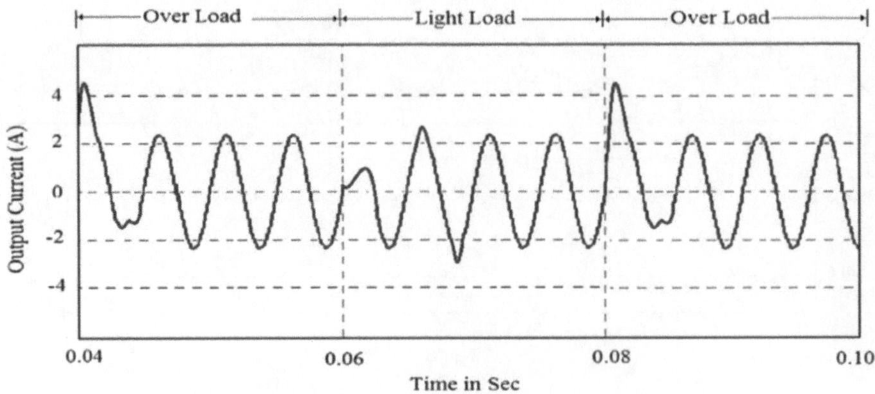

FIGURE 3.18 Inverter output current under load conditions (PID controller).

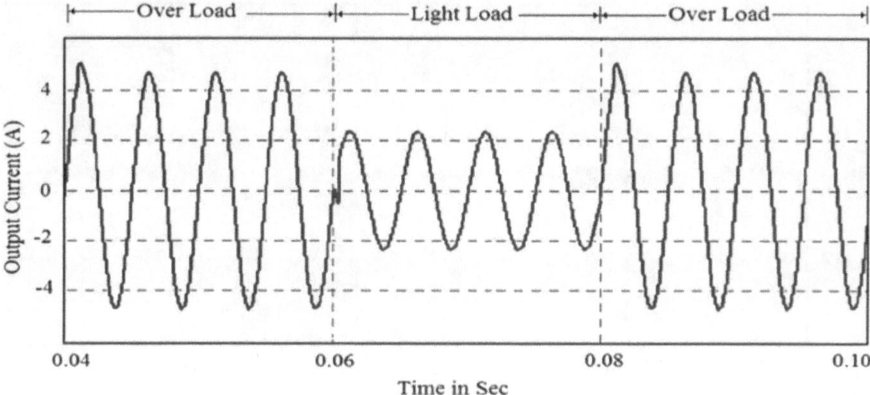

FIGURE 3.19 Grid output current under load conditions (FANN controller).

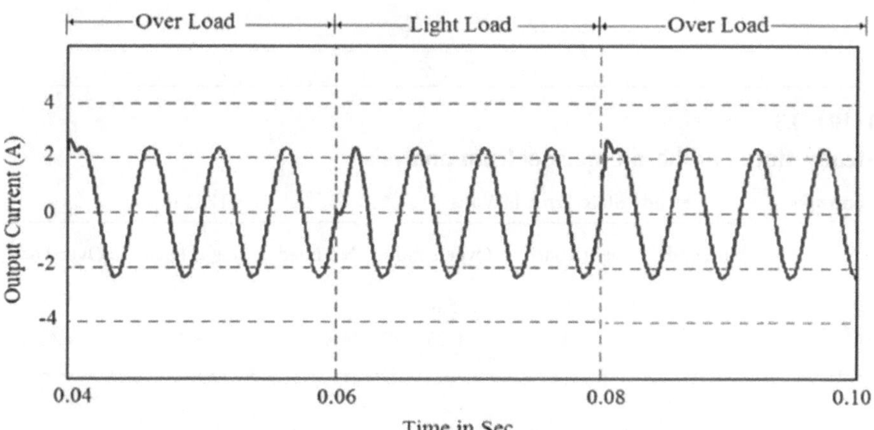

FIGURE 3.20 Inverter output current under load conditions (FANN controller).

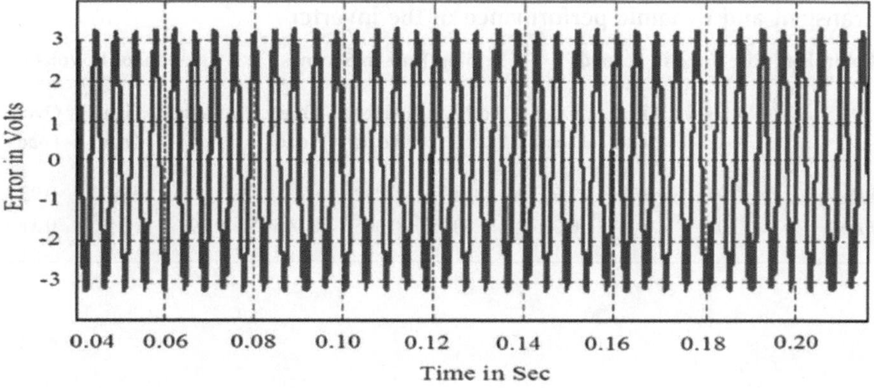

FIGURE 3.21 Steady state error for the inverter output voltage using PID controller.

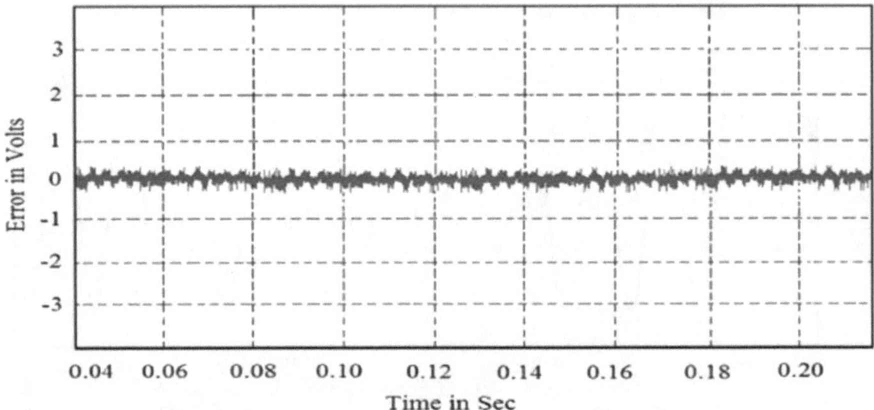

FIGURE 3.22 Steady state error for the inverter output voltage using FANN controller.

TABLE 3.3
Steady state and harmonic spectrum analysis

Controller	Steady State Error in Volts			THD in %		
	No Load	Light Load	Over Load	No Load	Light Load	Over Load
PID	5.5	5.2	5.6	9.6	10.5	11.6
FANN	0.2	0.23	0.25	2.0	2.4	2.42

TABLE 3.2
Transient and dynamic performance of the inverter

Controller	Rise Time in Seconds			Settling Time in Seconds			% Over Shoot In Volts		
	No Load	Light Load	Over Load	No Load	Light Load	Over Load	No Load	Light Load	Over Load
PID	0.07	0.08	0.092	0.08	0.092	0.1	0.8	0.82	0.93
FANN	0.02	0.03	0.035	0.005	0.005	0.005	0.49	0.5	0.51

FIGURE 3.23 (a) Experimental model with solar panel. (b) Assembled parts and recording of outputs using oscilloscope.

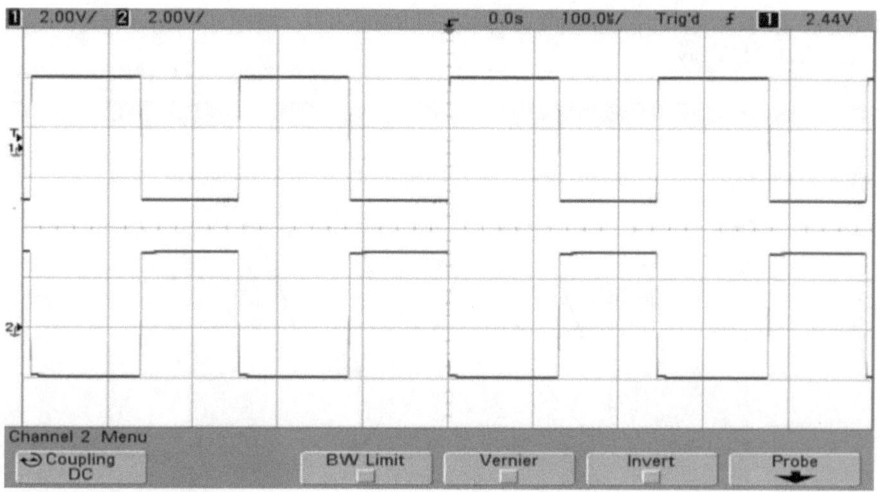

FIGURE 3.24 Driving pulses.

3.5.1 Hardware Implementation of Two Stage Single Phase LCL-T Inverter

Figure 3.25 shows that the waveform of current is sinusoidal and less harmonic which makes turn-off switching loss as reduced component.

The inverter current and voltage of the grid are in phase and has power factor = 1. The grid inductor current (L_G) and grid side capacitor voltage (V_{CF}) along grid are shown in Figures 3.25 and 3.26, respectively, using FFT spectrum analysis.

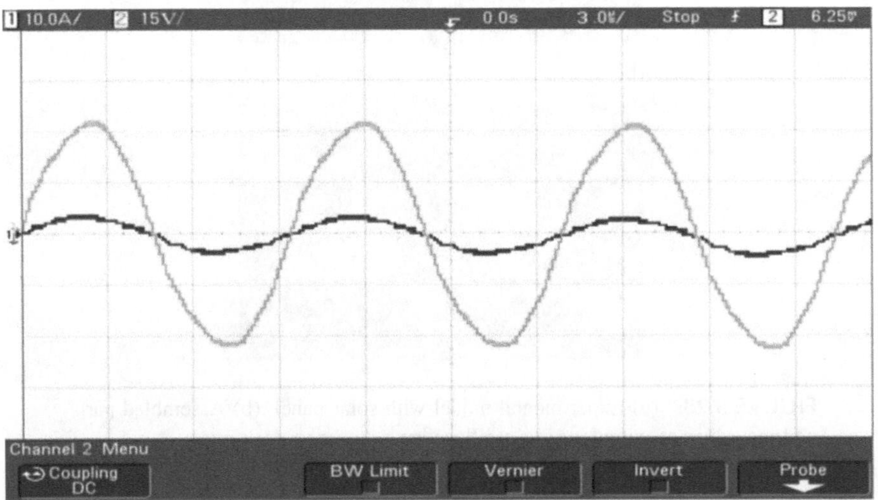

FIGURE 3.25 CH1: Grid current (Amp. Scale: 10A/div.), and CH2: Inverter output voltage (Volt. Scale: 15V/div.)

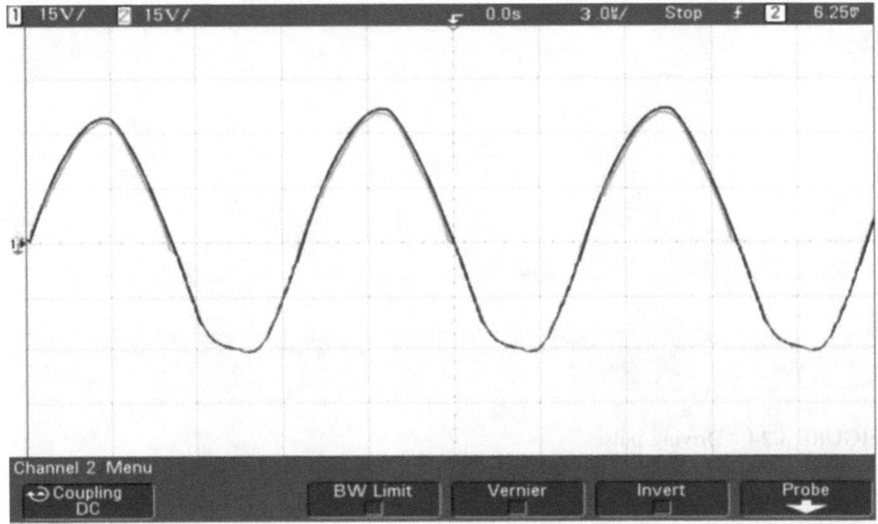

FIGURE 3.26 CH1: Grid voltage (Volt. Scale: 15V/div.), and CH2: inverter voltage (Volt. Scale: 15V/div.).

The inductor current waveform shown in Figure 3.27 is taken using Agilent Mixed Signal Storage Oscilloscope. The Figure 3.28 shows the FFT spectrum of the current across the inductor. It is clearly observed from Figure 3.29 the inductor current and capacitor voltage possess little oscillation and sinusoidal shape is improved. The grid side inductor current contains the harmonics of 3.22%.

The proposed FANN controller performance along output voltage of inverter with grid disturbance is shown in Figure 3.29. During the working of inverter, the grid contains 5 percentages of third order harmonics at 0.06 seconds. The settling time is 0.08s for closed loop system and overshoot is 0.84 V with better accuracy. Figure 3.30 shows the inverter output current with FFT analysis.

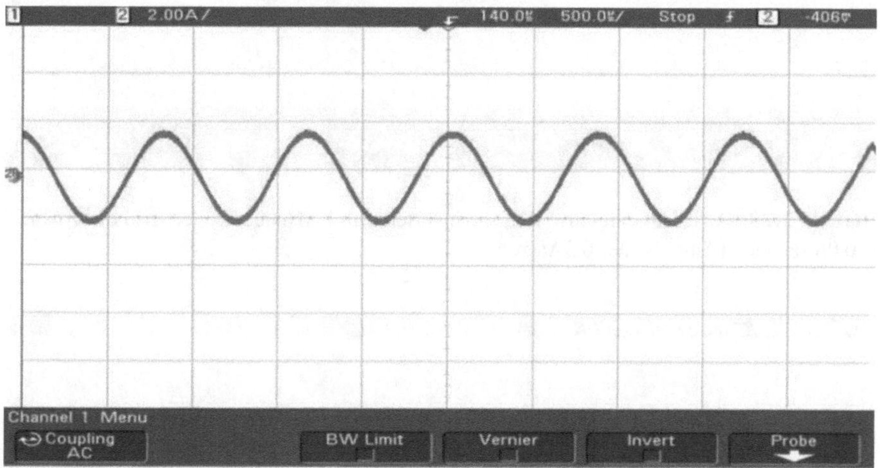

FIGURE 3.27 CH2: Grid inductor current (Amp. Scale: 2A/div.).

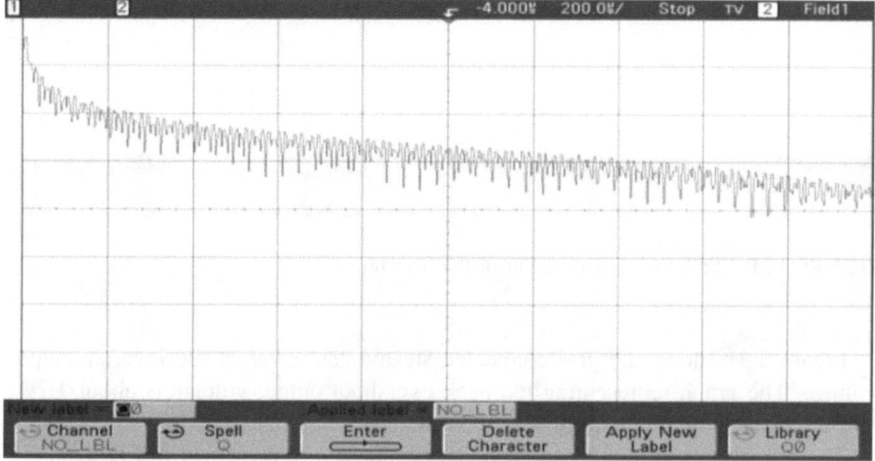

FIGURE 3.28 FFT for grid inductor current.

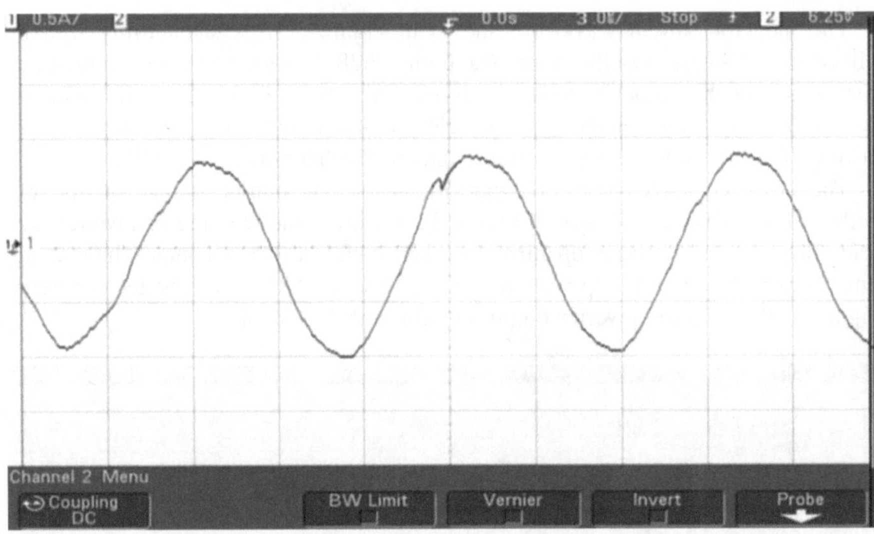

FIGURE 3.29 CH2: Output current of the inverter with FANN controller, grid disturbance at 0.06 seconds (Amp. Scale: 0.5A/div.).

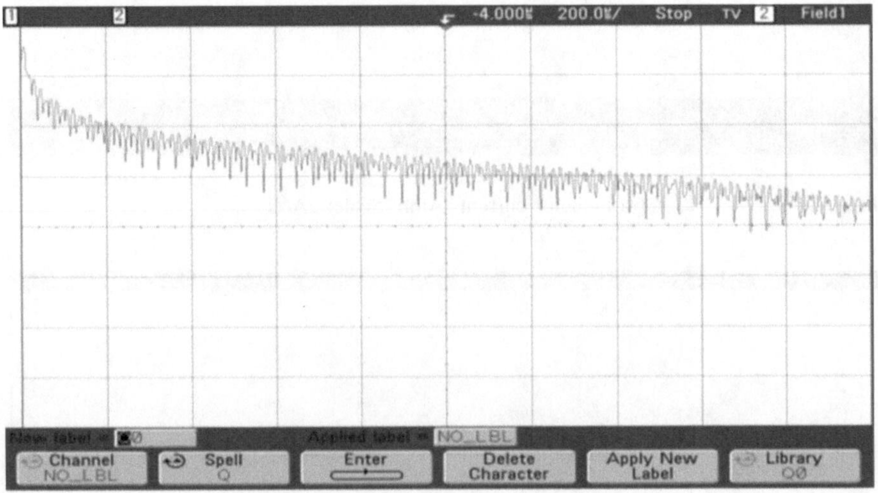

FIGURE 3.30 FFT for output current of the inverter.

Figure 3.31 shows the grid connected steady state error of the inverter output voltage. The graph representing the peak overshoot output voltage is about 1.2%, and the total harmonic distortion is about 4.25%.

The transformer primary and secondary voltages are presented in Figures 3.32 and 3.33, respectively. When voltage and current are minimum, power loss occurs during ON period and maintained low during inverter operation.

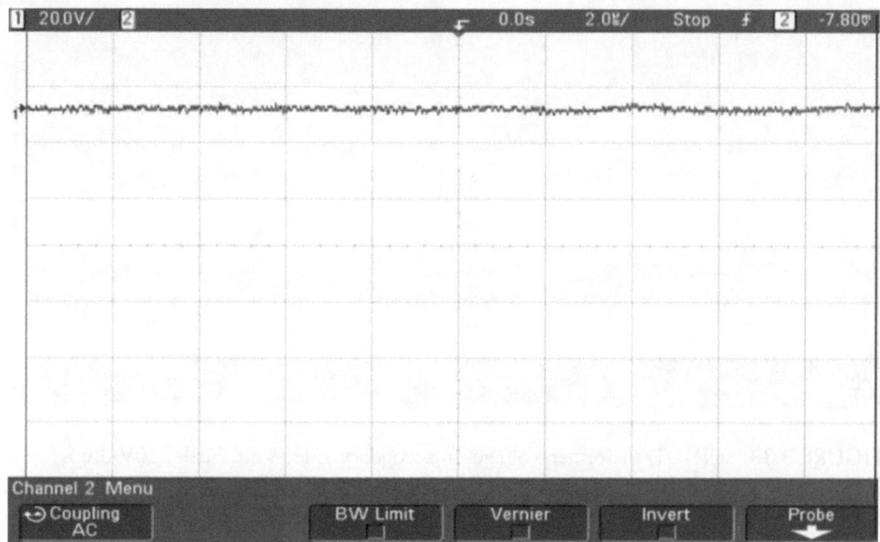

FIGURE 3.31 CH2: Steady state error for output voltage (Volt. Scale: 20V/div.).

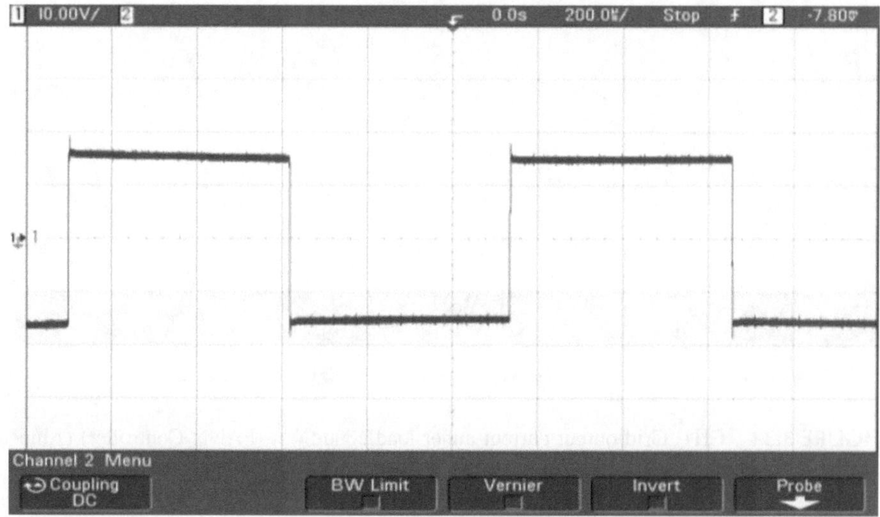

FIGURE 3.32 CH1: Transformer voltage on primary side (Volt. Scale: 10V/div.).

The grid inverter output current response under light load and over load conditions are shown in Figure 3.34 and Figure 3.35 respectively. The grid connected inverter operating under both conditions are tested with proposed FANN controller and presented in Figure 3.35.

A, 2 Ω resistive load is chosen for maximum load conditions. The oscillation and settling time of the current is better with minimum steady state error. The Table of

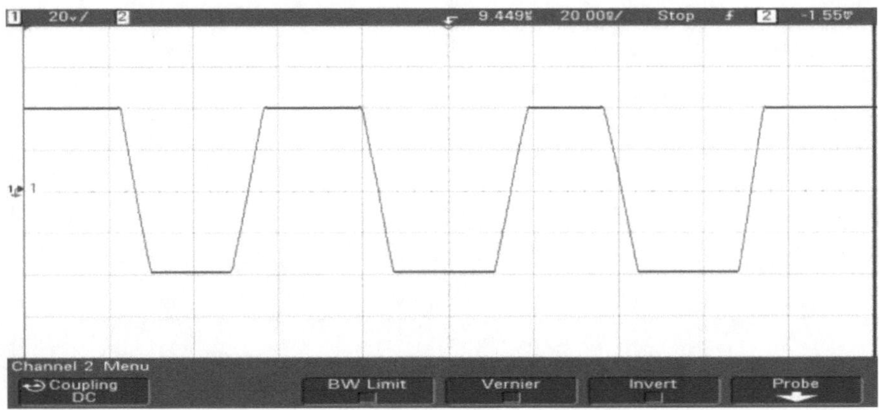

FIGURE 3.33 CH1: Transformer voltage on secondary side (Volt. Scale: 20V/div.).

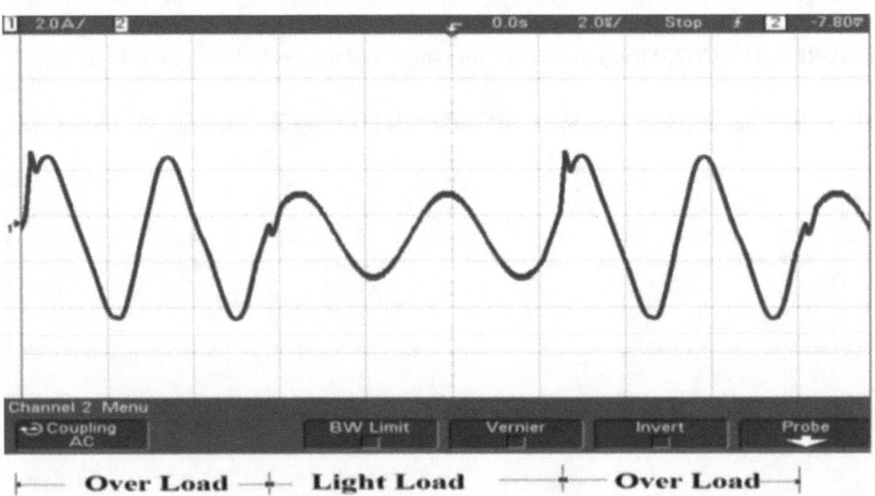

FIGURE 3.34 CH1: Grid output current under load conditions (FANN Controller) (Amp. Scale: 2.0A/div.).

LCL–T filter in Table 3.4 shows that the control characteristics match with theoretical values. The Table 3.5 shows the performance of the simulation and experimental results under various load conditions. The results prove the efficiency of the controller.

Table 3.5 representing elimination of percentage overshoot and THD is less using FANN control system based on LCL-T filter design under single phase grid connected mode.

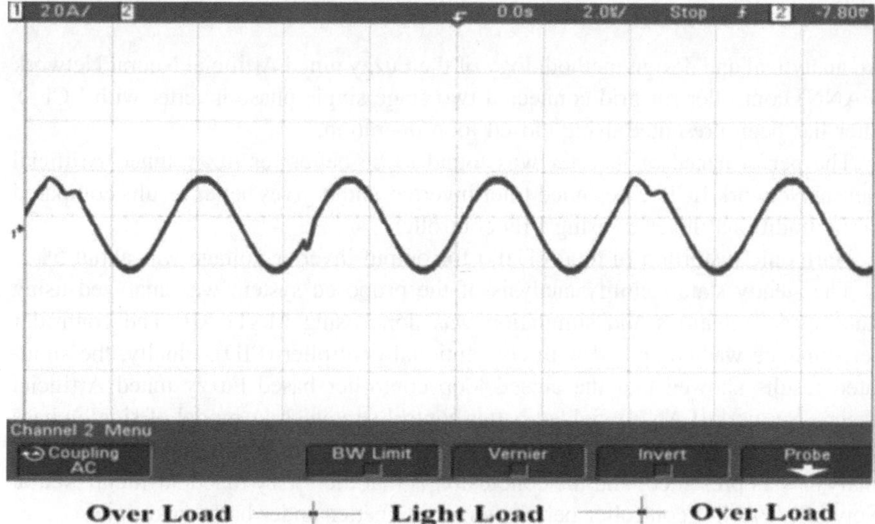

FIGURE 3.35 CH1: Inverter output current under load conditions (FANN Controller) (Amp. Scale: 2.0A/div.)

TABLE 3.4
Comparative analysis of the LCL-T filter inverter during grid connected mode

Performance Measures	Simulation Result		Experimental Results	
	Light Load	Over Load	Light Load	Over Load
Inverter Current in RMS	7.35	7.35	7.46	7.46
Grid Current in RMS	1.8	6.9	1.8	12.68
Load Current in RMS	5.4	13.9	5.68	6.25

TABLE 3.5
Performance measures of simulation and experimental results

Performance Measures	Simulation Results			Experimental Results		
	No Load	Light Load	Over Load	No Load	Light Load	Over Load
Settling Time in s	0.005	0.005	0.005	0.06	0.069	0.081
Rise Time in Sec	0.02	0.03	0.035	0.05	0.052	0.054
Percentage overshoot in Volt	0.49	0.5	0.51	0.72	0.84	0.89
Steady state error in Volt	0.2	0.23	0.25	0.89	0.9	1
THD	2.0	2.4	2.42	3.9	4.21	4.3

3.6 CONCLUSION

An analytical and design methodology of the Fuzzy tuned Artificial Neural Network (FANN) controller for grid connected two stage single phase inverter with LCL-T filter has been presented using closed loop operation.

The performance of inverter was found to be better for fuzzy tuned Artificial Neural Network LCL-T designed filter inverter, and it gives better results compared to the traditional inverter using PID controller.

Harmonic distortion in total (THD) for output inverter voltage was about 5%.

The steady state stability analysis of the proposed system was analyzed using state space equations and simulation was done using MATLAB. The controller performance was compared with conventional controller (PID). Finally, the simulated results showed that the closed loop controller based Fuzzy tuned Artificial Neural Network (FANN) yielded better control strategies. A model of single phase 300 W, 50 Hz inverter was designed and demonstrated. The harmonics spectrum analysis was presented, and the conclusion is that the Fuzzy tuned Artificial Neural Network (FANN) controller performance was better under both conditions.

REFERENCES

[1]. Chen Chen, Jian Xiong & Zhiqiang Wan 2017, 'A Time Delay Compensation Method Based on Area Equivalence For Active Damping of an LCL-Type Converter', *IEEE Transactions on Power Electronics*, vol. 32, no. 1, pp. 762–772.

[2]. Mahesh Kumar, Suresh Chandra Srivastava & Sri Niwas Singh 2015, 'Control Strategies of a DC Microgrid for Grid Connected and Islanded Operations', *IEEE Transactions on Smart Grid*, vol. 6, no. 4, pp. 1588–1601.

[3]. Claudio A. Busada, Sebastian Gomez Jorge & Jorge A Solsona 2015, 'Full-State Feedback Equivalent Controller for Active Damping in LCL-Filtered Grid-Connected Inverters Using a Reduced Number of Sensors', *IEEE Transactions on Industrial Electronics*, vol. 62, no 10, pp. 5993–6002.

[4]. Ammar Hussein Mutlag, Hussain Shareef, Azah Mohamed, Mohammed Abdul Hannan & Jamal Abd Ali 2014, 'An Improved Fuzzy Logic Controller Design for PV Inverters Utilizing Differential Search Optimization', *International Journal of Photo Energy*, vol. 2014, Article ID 469313, pp. 1–14.

[5]. Remus Narcis Beres, Xiongfei Wang, Marco Liserre, Frede Blaabjerg & Claus LethBak 2016, 'A Review of Passive Power Filters for Three-Phase Grid-Connected Voltage-Source Converters', *IEEE Journal of Emerging and Selected Topics in Power Electronics*, vol. 4, no. 1, pp. 54–69.

[6]. Xu Renzhong, Xia Lie, Zhang Junjun & Ding Jie 2013, 'Design and Research on the LCL Filter in Three-Phase PV Grid-Connected Inverters', *International Journal of Computer and Electrical Engineering*, vol. 5, no. 3, pp. 322–325.

[7]. Lin Chen, Ahmadreza Amirahmadi, Qian Zhang, Nasser Kutkut & Issa Batarseh 2014, 'Design and Implementation of Three-Phase Two-Stage Grid-Connected Module Integrated Converter', *IEEE Transactions on Power Electronics*, vol. 29, no. 8, pp. 3881–3892.

[8]. Paul Sergio Nascimento Filho, Tarcio Barros Dos Santos Barros, Marcos Villalva & E Ruppert Filho 2014, 'Design methodology of P-res controllers with harmonic compensation for three-phase DC-AC grid-tie inverters with LCL output filter', 2014 IEEE 15th Workshop on Control and Modeling for Power Electronics (COMPEL), Santander-Spain.

[9]. N Sandeep, Prakash S Kulkarni & R Yargathi Udaykumar 2014, 'A single-stage active damped LCL-filter-based grid-connected photovoltaic inverter with maximum power point tracking', IEEE Conference Publications, Eighteenth National Power Systems Conference (NPSC), Vadadora, pp. 1–6.

[10]. Aurobinda Panda, MK Pathak & SP Srivastava 2016, 'A Single Phase Photovoltaic Inverter Control for Grid Connected System', *Indian Academy of Sciences*, vol. 41, no. 1, pp. 15–30.

[11]. Costantino Citro, Pierluigi Siano & Carlo Cecati 2016, 'Designing Inverters' Current Controllers with Resonance Frequencies Cancellation', *IEEE Transactions on Industrial Electronics*, vol. 63, no. 5, pp. 3072–3080.

[12]. Perumala Chandrasekhar & Subham Rama Reddy 2010, 'Design of LCL Resonant Converter for Electrolyser', *International Journal of Electronic Engineering Research*, vol. 2, no. 3, pp. 435–444.

[13]. Behrooz Mirafzal, Mahdi Saghaleini & Ali Kashefi Kaviani 2011, 'An SVPWM-Based Switching Pattern for Stand-Alone and Grid-Connected Three-Phase Single-Stage Boost-Inverter', *IEEE Transactions on Power Electronics*, vol. 26, no. 4, pp. 1102–1111.

[14]. Zhen Qin, Deshang, Sha & Xuehua Liao 2012, 'A three-phase boost-type grid connected inverter based on synchronous reference frame control', 27th Annual IEEE Applied Power Electronics Conference & Exposition, USA, pp. 384–388.

[15]. Mahdi Saghaleini & Behrooz Mirafzal 2012, 'Reactive power control in three-phase grid connected current source boost inverter', 27th Annual IEEE Applied Power Electronics Conference & Exposition, USA, pp. 904–910.

[16]. Jiabing Hu & ZQ Zhu 2011, 'Investigation on Switching Patterns of Direct Power Control Strategies for Grid-Connected DC-AC Converters Based on Power Variation Rates', *IEEE Transactions on Power Electronics*, vol. 26, no. 12, pp. 3582–3598.

[17]. Akanksha Singh & Behrooz Mirafzal 2019, 'An Efficient Grid-Connected Three-Phase Single-Stage Boost Current Source Inverter', *IEEE Power and Energy Technology Systems Journal*, vol. 6, no. 3, pp. 142–151.

[18]. Naoto Kobayashi, Yuji Hayashi, Seiji Iyasu & Yuichi Handa 2019, 'Fast current control of the single-phase DC-AC converter using digital peak current mode control', 2019 21st European Conference on Power Electronics and Applications (IEEE), Genova, Italy.

4 Motion Vector Analysis Using Machine Learning Models to Identify Lung Damages for COVID-19 Patients

Dr. Malik Mohamed Umar[1], *Murugaiya Ramashini*[2], *and M.G.M. Milani*[3]

[1]School of Engineering and Applied Sciences, Kampala International University

[2]Department of Computer Science and Informatics, Faculty of Applied Sciences, Uva Wellassa University

[3]Faculty of Integrated Technologies, Universiti Brunei Darussalam

CONTENTS

DOI: 10.1201/9781003194415-4

4.1 INTRODUCTION

An estimated 334 million people who are exposed to viruses in the air may develop respiratory diseases that could cause severe damages to the lungs and other organs of the respiratory system. The lungs are one of the most critical organs in the body that allow the flow of oxygen to our body and the flow out of carbon dioxide that keep us healthy and capable of normal work and life. Numerous lung diseases may preclude the intake of enough oxygen to the body and may cause severe damages to other organs, as well as to the blood circulation. Lung diseases can happen due to problems in the airways, lung tissues and the blood vessels. Some lung disorders or respiratory failures may have origin from birth, and those diseases are called chronic lung diseases [1]. The most common and severe chronic lung diseases that occur in the airways are asthma, chronic bronchitis and emphysema. These diseases can cause severe damages to the lungs over time. During our breathing, the lung tissues and muscles help to expand the lungs like a balloon. However, if there is inflammation of the tissues, then the lungs may fail to fully expand, and the lungs will have to put in a high effort to deliver the oxygen to the blood and expel the carbon dioxide from the body.

Many other lung syndromes affect the blood vessels in the lungs, such as scarring, clotting or inflammation of the blood vessels [2]. The lung circulation diseases can cause critical damages to the functionality of the heart and brain that later cause death. Many lung diseases are combined with the various disorders of the airways, lung tissues and lung circulation. The most common lung diseases are: swelling and inflammation of bronchial tubes, Chronic Obstructive Pulmonary Disease (COPD), pneumonia, lung cancer, blockages in the arteries, and abnormal fluid build ups in the lungs [2]. The detection of these respiratory diseases is widely examined by medical video and image analyzing tools. Among many available medical examination tools, the ultrasound video analysis plays a crucial role in supplementing the varieties of information on lungs' functional behaviour to detect any present respiratory disease. The combination of traditional methods on the lung disease diagnostic process and modern technology may have a higher impact by detecting the diseases when still in milder stages. Thus, machine learning can be defined as the most influential technological approach in the medical field. The varieties of machine learning techniques give many opportunities to select the best method to assess the best results since they offer significant advantages in health care data analytics and evaluation tasks.

The integration of machine learning algorithms in the healthcare system may deploy many flexible and scalable solutions in classifying and predicting different types of diseases for early diagnosis such as medical image analysis, video analysis, sound analysis or laboratory findings. Manogaran et al. [3] used Orthogonal Gamma Distribution with a Machine Learning Approach (OGDMLA) to evaluate the Magnetic Resonance Imaging (MRI) images to detect brain tumors. Milani et al. [4] identified the normal and abnormal heart sounds using Principal Component Analysis (PCA) along with Artificial Neural Networks (ANN), while Cruz et al. [5] applied ANN for cancer prediction. Verma et al. [6] introduced the k-Nearest Neighbor (kNN) algorithm to identify the kidney stones by analyzing the ultrasound images of the kidneys. Roy chowdhury et al. [7] analyzed the fundus images to identify the severity grade for diabetic retinopathy (DR) using Gaussian Mixture Model (GMM), Support Vector Machine (SVM), kNN, and AdaBoost. [8] applied Classification Tree (CT) machine learning algorithm to predict the increased blood pressure by body mass index (BMI). D'Souza et al. [9] applied Decision Tree (DT) and SVM machine learning models to differentiate HPV-positive and HPV-negative patients by Head and Neck Squamous Cell Carcinoma (HNSCC). Zhang et al. [10] applied Convolutional Neural Networks (CNN) to detect lung tumors from the Positron Emission Tomography (PET) images. Kong et al. [11] applied cascade CNN to detect thyroid nodules in ultrasound images.

Moreover, Santos et al. [12] identified small lung nodules in 3D computerized tomography scan (CT) data by applying GMM and SVM. Lin et al. [13] identified shoulder pains using the ultrasound images by applying CNN. Correspondingly, Isitha et al. [14] applied CNN to determine the lung diseases by analysing texture patterns of CT images. Among all types of healthcare data, the images and videos are the top rewarding resources in medical data analysis. Many clinicians rely on medical image and video analysis to estimate the specific disease of the patients and monitor treatment to improve outcomes. Consequently, medical images and videos are the largest source of data in the healthcare system that includes much worthwhile information.

Ultrasound is the safest and cost-effective medical imaging method that is available to all. Ultrasound imaging contains varieties of applications (i.e. cardiac diagnosis, diagnosis of the spine and internal organ). Medical ultrasound images use high-frequency sound waves to capture photos of parts of the inside of the body. The ultrasound videos are produced by combining the real-time images. Thus, it can display different portions of parts of the body such as organs, vessels and tissues in one frame to allow the clinicians to have a better understanding of the patient's condition and to assign a particular treatment [15]. Apart from many other organs, ultrasound is commonly used to observe the functional behaviour of the lungs. The popularity of ultrasound scanning in lungs has increased because it carries sufficient information to diagnose the common lung pathologies such as pleural pathology, pericardial pathology, shortness of breath, cyanosis, cough, and shock [16].

Nowadays, many machine learning techniques are integrated with ultrasound videos to achieve better predictions of the functional behaviors of the lungs. However, from the viewpoint of both ultrasound lung video and image analysis, it is vital to develop accurate automatic lung disease and respiratory system disorder classification methods using modern technology. Especially, using advanced machine learning and deep learning algorithms with general image and video processing techniques have been investigated by [17]. Correa et al. [18] applied ANN to differentiate the ultrasound images of pneumonia infected lungs from the healthy lungs of young children and achieved 90.9% sensitivity and 100% specificity. Zhou et al. [19] developed a Lung Ultrasound Surface Wave Elastography (LUSWE) technique by integrating Deep Neural Network (DNN) model to analyse the mass of the lung tissue and achieved 99.2% of validation accuracy. This proposed DNN model was further improved by Zhou et al. [20] to analyse the Lung Mass Density (LMD) of Interstitial Lung Disease (ILD) in patient's ultrasounds and achieving

89% than those of machine learning models. Palacio et al. [21] proposed an imaging biomarker-based Automatic Quantitative Ultrasound Analysis (AQUA) texture extractor to evaluate fetal lung maturity and achieved 90.3% accuracy.

Similarly, Prakash et al. [22] and La Torre et al. [23] assessed the fetal lung maturity using the ultrasound images and achieved classification accuracies in the range of 73% to 96%. Carne et al. [24] analyzed the texture of ultrasound images of the fetal lungs using machine learning models such as regression, classification trees and neural networks. They achieved 86.2% and 87% as the sensitivity and specificity, respectively. Yanran et al. [25] analyzed fetal lung texture of lung ultrasound images by applying regression machine learning model and achieved over 75% classification accuracy. Mehanian et al. [26] proposed a CNN based Pneumothorax (PTX) detection using lung ultrasound videos. The videos were fed into the proposed classification models without preprocessing and achieved over 83% accuracy. Cheng et al. [27] categorized eleven (11) categories of abnormal ultrasound images using CNN and achieved 77.9% accuracy, while Kulhare et al. [28] applied the same model to detect five (05) lung abnormalities in ultrasound images. Than et al. [29] adopted SVM classifier to classify Riesz-based features of the lung images to identify different types of lung diseases and achieved 99.53% classification accuracy. Burgos et al. [30] extracted the Empirical Mode Decomposition (EMD) features from the ultrasound images. Dimensionality of the extracted EMD features was reduced using the PCA algorithm to classify the healthy and pneumonia patients' lungs using the kNN machine learning model and achieved 83.33% accuracy. Similarly, Prasanth et al. [31] applied the kNN classification model to detect the changes in lung fluids and achieved 91.7% accuracy. Veeramani et al. [32] extracted features by the convoluted local tetra pattern, the Haralick feature extraction, the histogram of oriented gradient and the complete local binary pattern features from the lung ultrasound images. Extracted features were classified using multi-level Relevance Vector Machine (RVM) classifier to identify lung diseases such as Respiratory Distress Syndrome (RDS), transient tachypnea of the newborn, pneumothorax, pneumonia, bronchiolitis, meconium aspiration syndrome, and lung cancer. This proposed classification method achieved 100% accuracy.

The research studies which applied regression, classification tree and CNN models achieved less than 90% classification accuracy [24,27], while many other research studies achieved over 90% classification accuracies using ANN, DNN, SVM and kNN models. The difference between these classification results shows that the classification model selection is performing a significant role in ultrasound video and image analyzing tasks. It could be stated that the effect of each classification model may provide various classification results depending on the feature selection made. This statement was well proved in [32] which achieved 100% classification accuracy for their extracted features. Therefore, it is vital to select the best features before the selection of machine learning classification models to classify normal vis-a-vis abnormal images.

4.1.1 Background of the Study

Recently, the lung ultrasound video analysis surpassed the other ultrasound analyzing applications and took a leading role in many medical findings. This attention received by ultrasound video analysis may have been created by the COVID-19 pandemic, which has triggered a potential health threat in countries worldwide. COVID-19 disease is caused from the Severe Acute Respiratory Syndrome Coronavirus 2 (SARS-CoV-2) virus. The SARS-CoV-2 is enveloped, highly diverse and single-stranded RNA virus that originated from bats [33]. As the name implies, the patients with SARS-CoV-2 are afflicted by respiratory diseases. According to the World Health Organization (WHO), these kinds of viruses attack the lungs in three stages: at first, the virus spreads to an individual from an infected person, and viral replication takes place. After that, the virus infects and kills the cilia cells, where cilia is the hair-like structure present in the lungs to protect it from pathogen entry.

Damage in the ciliary cells makes the lung fill up with debris and fluids, which leads to shortness of breath, and it is the primary symptom developed by COVID-19 patients. In order to repair the damaged lung and tissues, the immune cells get flooded into the lungs. Then, they start to invade the virus. Nevertheless, sometimes hypersensitivity takes place, in which the immune cells damage the healthy cells, and this condition is referred to as the immune hyper-reactivity, which is the second stage of the viral attack. Continuous lung damage will result in a third stage, namely, pulmonary destruction in which the respiratory organ fails to work, which leads to death. Not all patients with COVID-19 are likely to endure all three stages; only 25% of patients encountered pulmonary destruction, and many of them will survive with chronic lung damage [34].

4.1.2 Motivation and Problem Statement

In general, viruses and many other diseases will affect the breathing functionality of the lungs. Among many other lung diseases, COVID-19 has caused severe damages globally from trying to stem its spread that has affected healthcare, finance, industry and education. Reports from the WHO show that there are more than 17,900,500 confirmed COVID-19 patients worldwide, including over 680,000 deaths [35]. Furthermore, the WHO states that many vulnerable people have a high chance to catch this virus and which can later cause perilous damages [36]. An effective treatment is highly in demand to treat this virus. However, the current focus of the researchers is to find antivirals or a vaccine. Nevertheless, not a few research studies are conducting their studies on analyzing the ultrasound videos of lungs for early prediction of COVID-19 infected patients to control the spread of the virus while waiting for the pending vaccine treatment.

This study analyzes the inspiration (inhalation) and expiration (exhalation) patterns of lungs to identify the specific changes of COVID-19 infected patients. For this analysis, the ultrasound videos of lungs are taken into consideration because they can provide many lung movement information that can be taken in with just a single glance of the eyes. The motion that happens during breathing will be taken into consideration in this study to identify the changing properties of the lung movements. The Non-Rigid Bodies (NRB) algorithm is adopted in this work to find the changing lung movement properties by computing the motion vectors of subsequent frames in the lung video sequences. A few supervised machine learning classification models are proposed in this work for further analysis of computed motion vectors.

4.1.3 Structure of the Chapter

The rest of the paper is organised as follows: **Section 2** describes the proposed methodology to achieve the defined problem statement. The proposed methodology is divided into four subsections: data collection and preprocessing, feature extraction, feature processing, and classification. **Section 3** presents results and discussions of the feature analysis and classification. **Section 4** provides the final remakes and the conclusions of the research study. Moreover, this section presents the possible future directions for expansion and development of this work.

4.2 PROPOSED METHODOLOGY

The necessary steps of the proposed lung ultrasound video classification methodology are shown in Figure 4.1. The input ultrasound videos were collected and preprocessed before being divided between training and testing datasets. Then, features were extracted from both training and testing datasets. The extracted features were further processed before the classification task was accomplished. Finally, the supervised machine learning approach was implemented to predict the COVID-19 infected people from the normal people.

4.2.1 Data Collection and Pre-Processing

The dataset used in this study is a publicly available online database which includes lung ultrasound videos of normal subjects and patients who are infected with COVID-19 virus, bacterial pneumonia or viral pneumonia. The dataset is updated regularly by the authors and contributors. However, we have chosen the most recent dataset which was updated in July 2020. More details about the dataset can be found at:

https://github.com/jannisborn/covid19_pocus_ultrasound/tree/master/data

From the database, it has been noticed that five (05) COVID-19 lung ultrasound videos and another five (05) normal lung ultrasound videos have visible patterns that may be incorporated with our proposed method. Then, all the selected ten (10) ultrasound video datasets (i.e. five (5) COVID-19 videos and five (05) normal subjects' videos) are assembled to execute the artefacts by manually cropping the videos. However, the selected dataset of this study is less. The manually cropped videos are further preprocessed by applying linear Despeckle filters to remove the speckle noise [37,38]. The speckle noise is a common grainy salt and pepper pattern noise available in medical ultrasound videos, which mask the crucial details.

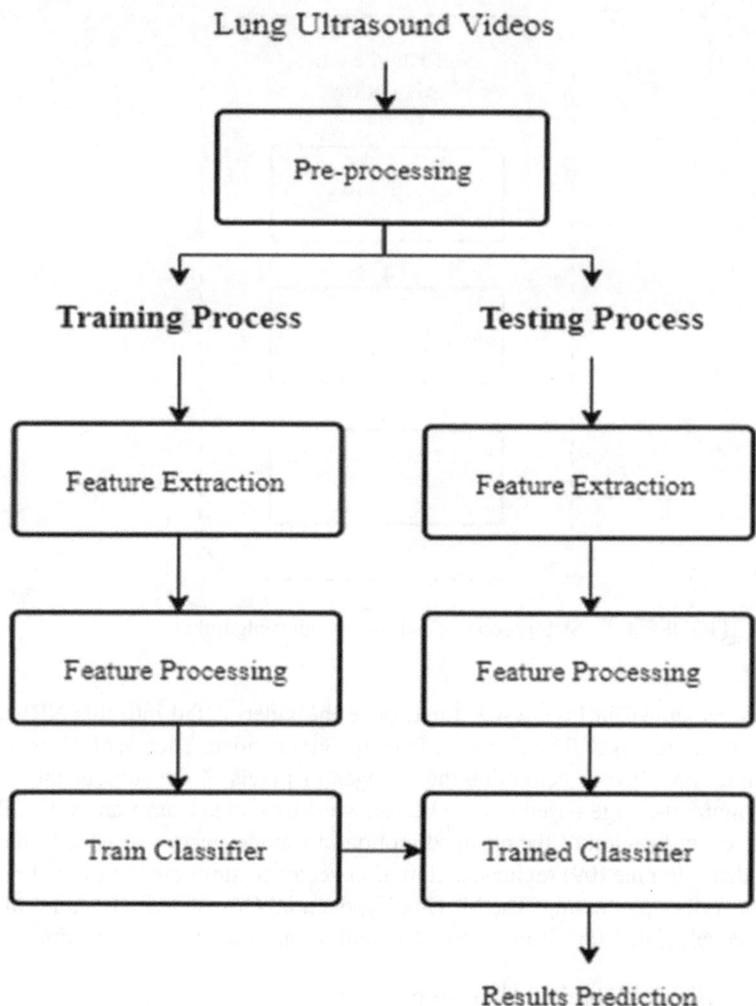

FIGURE 4.1 Block diagram of the proposed methodology.

Finally, the preprocessed ultrasound videos are divided as training and testing datasets for feature extraction and classification. The training dataset includes ultrasound videos of three (03) COVID-19 subjects and three (03) normal subjects. The testing dataset includes two (02) COVID-19 subjects' videos and two (02) normal subjects' videos.

4.2.2 Feature Extraction

The NRB algorithm is adopted to extract the motion vector values from the lungs' ultrasound videos. This algorithm includes a region-based adaptive search and Structure from Motion (SFM) techniques to extract edge factor-based motion vector

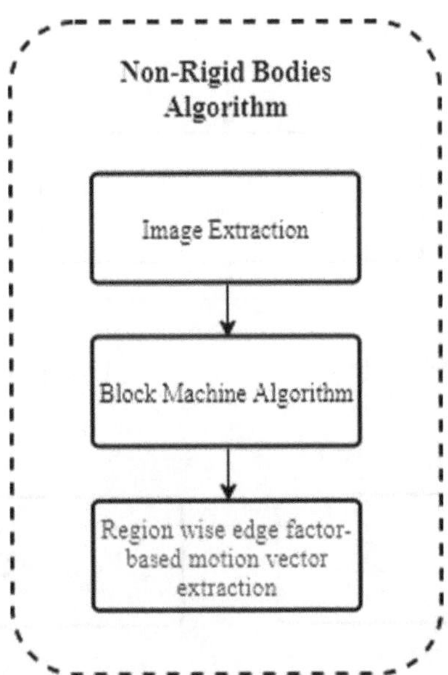

FIGURE 4.2 Step process of non-rigid bodies algorithm.

features, as shown in Figure 4.2. First, once the video is fed into this NRB algorithm, image frames will be extracted from the given video. Then, a block-matching algorithm is applied to determine the changes of pixels of the subsequent frames. Furthermore, the edge-based motion vectors were extracted from frames. Finally, to find the exact location of the changed motion in frames, every extracted frame will be divided into nine (09) regions, and motion vector features are considered region-wise for further processing. The Moving Exponential Growth (MEG) values of each region is calculated for all the extracted frames and used for classification.

4.2.2.1 Block Matching Algorithm

The block matching algorithms were used to evaluate the motion in the frames of video sequences. In these algorithms, each frame is divided into groups of neighbouring pixels called macroblocks. The overall goal is to locate each macroblock of the current frame for a candidate block in reference frame by exploring the displacement of motion between the blocks. Each displacement will produce a motion vector of a current block in the current frame by considering its position in the reference frame. In simple words, this algorithm can search and locate similar macroblocks between frames. Various search algorithms were adopted based on the applications to fulfil the above task.

The Adaptive Rood Pattern Search (ARPS) is one of the commonly used block matching algorithms [39]. Nie et al. [40] introduced the ARSP as a simple and fastest block-matching motion estimation algorithm which has two stages, namely: initial search and refined local search. The size of the Adaptive Rood Search (ARS) for each macroblock is dynamically determined based on the available motion vectors of neighboring blocks in the first stage. At the same time, a Unit-size Rood Pattern (URP) is exploited unrestrictedly and repeatedly until the final macroblock is found.

To befit different motion contents of the video sequence for each macroblock, the ARSP will exploit a higher distribution of macro vectors in both horizontal and vertical directions, alongwith spatial inter-block correlation. Further, it will exploit the adjustable rood-shaped search pattern with a search point indicated by the predicted motion vector, which is shown by:

$$\Gamma = Round|\overrightarrow{MV_{predicted}}| \tag{4.1}$$

$$\Gamma = Round\left[\sqrt{MV^2_{predicted}(x) + MV^2_{predicted}(y)}\right] \tag{4.2}$$

$$\Gamma = Max\{|MV_{predicted}(x)|, |MV_{predicted}(y)|\} \tag{4.3}$$

where, $MV_{predicted}(x)$ is horizontal component of Motion Vector (MV), and $MV_{predicted}(y)$ is vertical component of the Motion Vector (MV) [40].

The search speed and quality of the performance will be influenced by the shape and size of the search [41]. Through a performance-based study on block matching algorithms, Choudhury et al. [39] concluded that ARPS algorithm is more economical and has good prediction quality compared to other search algorithms for motion estimation with very low complexity. Therefore, the ARSP is adopted as a block matching algorithm in this work.

4.2.2.2 Region-Wise Edge Factor-Based Motion Vector Extraction

The edge factor-based detection is used to identify the non-rigid boundary of the object. Edge factor-based detection aims to detect the changes, from the structure which could be evaluated by the motion pixels. The region-wise edge factor-based motion vectors were extracted to get a deeper understanding of the changes in functionalities of the lungs [42,43].

Due to the virus infection, inspiration and expiration patterns of the lungs will be affected. Thus, the region-wise edge factor-based motion vectors were extracted using the NRB algorithm to detect the non-rigid structure changes of lungs. The frame of the ultrasound video sequence is divided into nine regions. The proposed region-wise separation is shown in Figure 4.3.

4.2.3 Feature Processing

The feature processing is proposed to analyze and understand the extracted features before further classification. To find the increase of the quantity of lung tissue over time, the exponential growth statistical function of previously extracted motion vectors is calculated by:

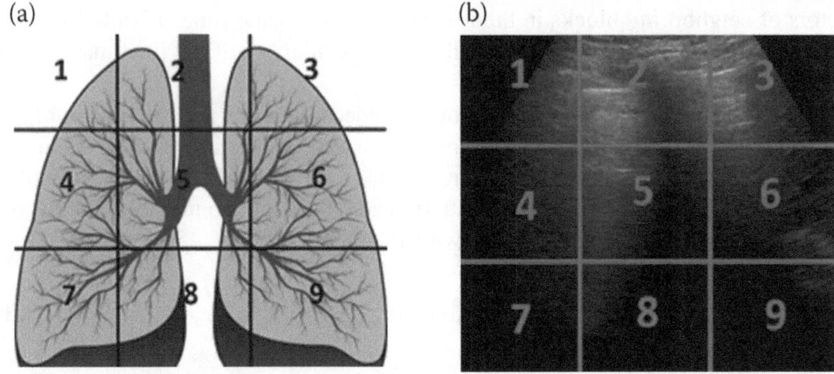

FIGURE 4.3 Region-wise separation of a human lung (a) actual lungs (b) corresponding ultrasound image.

$$y = a(1 + r)^x \tag{4.4}$$

where, a is extracted motion vector value (prior to the measuring growth), r is growth rate, x is number of time intervals that have passed [44].

Then the PCA is applied to the calculated exponential growth values (y) to reduce the input features and identify the most important features that may provide better classification results in the classification process. Before applying the PCA, the feature matrix is formed as $m \times n$, where the m rows are the lung ultrasound video samples, and n columns are the calculated exponential growth values. Then the PCA of $m \times n$ feature matrix is calculated by:

$$Y = P(\mathrm{X}) \tag{4.5}$$

where, X is $m \times n$ feature matrix, Y is transformed matrix, and P is $m \times n$ projection matrix.

The main advantage of using PCA in this proposed method is its ability to transmit the most relevant features in descending order [45]. The first column of the transformed matrix (Y) consists of the principal components which have the maximal amount of variance. Subsequently, the first three (03) principal components are input to the supervised classification models at the classification process to achieve better classification results.

4.2.4 Classification

The principal idea of this study is to build an advanced lung ultrasound video prediction model using machine learning algorithms to differentiate normal and COVID-19 infected lungs. The proposed classification process is constructed using six (06) popular supervised machine learning algorithms such as DT, Logistic Regression (LR), Linear Discriminant Analysis (LDA), SVM, kNN and RT; yet each classification model iterates separately to classify the first three (03) PCA values of normal and COVID-19 infected lung ultrasound videos.

The main reasons for selecting the above six (06) classifiers are their simplicity and high efficiency, though the supervised machine learning algorithms use pre-labelled data to train the classification models to predict the output of any unknown data. The DT classifier creates a tree architecture and computes the outcome at the top of the tree by considering one feature at a time and grows many small tree branches to classify each remaining feature, whilst Random Forest (RF) model selects the variables randomly to build multiple decision trees [46]. The LR and LDA methods are popular linear classification models that rely on the linear-odd assumptions to estimate the coefficients [47]. The SVM adopts nonlinear mapping to separate the input data linearly to execute binary data classification tasks [48]. The kNN is a statistical classification model that calculates the average response of the variables from the k samples of calibration data to find the labels of unknown samples [49]. All these six (06) proposed supervised machine learning algorithms learn the mapping function of the input training feature matrix and output the trained feature matrix by:

$$Y_c = F(Y) \tag{4.6}$$

where, Y_c is a trained feature matrix (trained model), f is mapping function and Y is the transformed feature matrix. Finally, the testing feature matrix is sent to the trained model (Y_c) to predict which class belongs to them (i.e. COVID-19 subjects and normal subjects).

4.3 RESULTS AND DISCUSSION

In this study, two (02) types of analysis have been carried out; type one (01) is feature analysis, and type two (02) is classification-based analysis. As depicted in Figure 4.1, after applying NRB algorithm to the collected ultrasound video sequences, the extracted features have been analyzed to get a better understanding on differentiating features of COVID-19 infected ultrasound lung videos over normal ultrasound lung videos before using machine learning classification-based identification. Further analysis has been extended with the prediction results of six (06) supervised machine learning classifiers.

4.3.1 Feature Analysis

Once region-wise edge factor-based motion vectors of each frame are extracted using the NRB algorithm, the features of randomly selected one (01) COVID-19 infected as well as one (01) randomly selected normal ultrasound video have been plotted. This analysis aims to identify the functional changes of lungs in deferent regions. Figure 4.4 illustrates the region-wise motion vectors of COVID-19 infected lung video and Figure 4.5 illustrates the region-wise motion vectors of normal subject lung video.

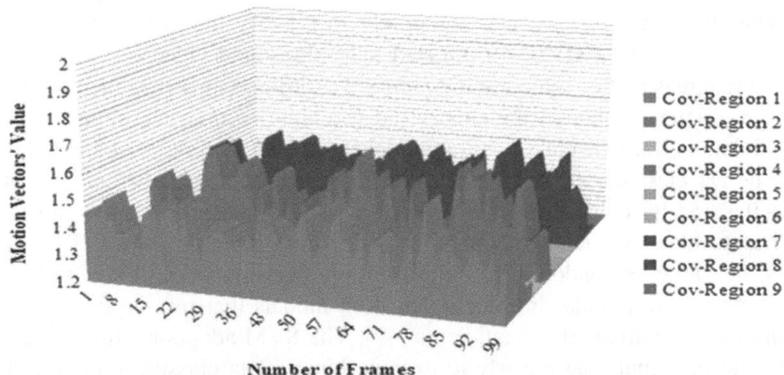

FIGURE 4.4 Motion vector value of a randomly selected COVID-19 infected lung ultrasound video.

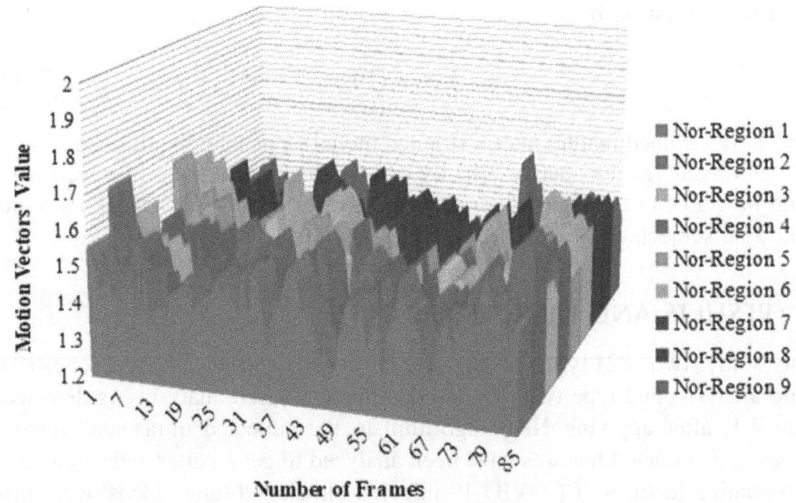

FIGURE 4.5 Motion vector value of a randomly selected normal lung ultrasound video.

As shown in Figure 4.4, the region four (04) and region seven (07) have vital movements compared to other regions in COVID-19 infected lungs, whereas in the normal subject, the lung movements that are shown in Figure 4.5 are competent in all regions. Both Figures 4.4 and 4.5 show that the region one (01) has a little movement. However, it does not qualify as an adequate movement to make a significant difference in COVID-19 infected lung and normal subject lung videos. Thus, the results prove the fact of normal subject lungs as always flexible (non-rigid). According to the region-wise feature analysis examination, it shows that the COVID-19 infected lungs may be of reduced flexibility by affecting their movements. Moreover, the MEG of four (04) normal subject lung videos is compared with a COVID-19 infected lung video to find the differences between the

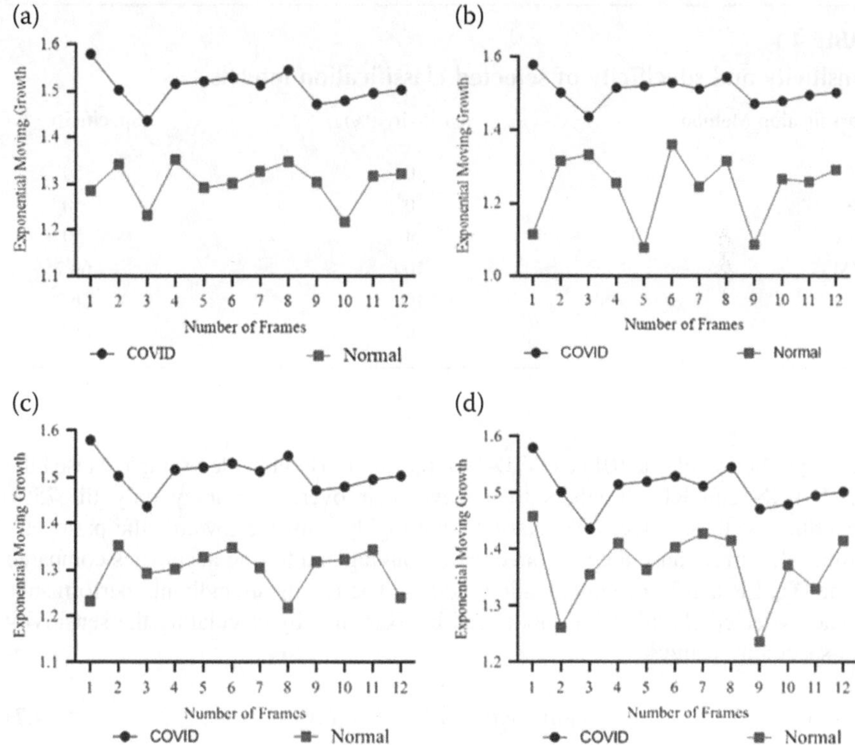

FIGURE 4.6 (a, b, c, d): Comparison of one (01) COVID-19 infected ultrasound video with four (04) different normal lungs ultrasound videos.

COVID-19 infected lungs and normal lungs ultrasound videos. Figure 4.6. shows the comparison results between COVID-19 infected lung and normal lung ultrasound videos.

Overall, feature analysis results of this section demonstrate that the region-wise edge factor-based motion vectors which have been extracted using NRB algorithm can differentiate COVID-19 infected and normal ultrasound videos. Therefore, the calculated MEG of region-wise edge factor-based motion vectors have been fed to selected supervised machine learning classifiers and the classification results are elaborated in the following section.

4.3.2 Classification Analysis

The classification accuracy can be utilized to measure the engagement of the selected classifiers (i.e. DT, LR, LDA, SVM, kNN and RF) in classifying COVID-19 infected lungs and normal lungs. The test accuracy of each selected classifier depends only on the region-wise edge factor-based motion vectors of the selected four (04) lung ultrasound video samples (i.e. two (02) COVID-19 infected, and two (02) normal). Among all six (06) classifiers, the DT, LR and LDA have 50% overall accuracy. The

TABLE 4.1
Sensitivity and specificity of selected classification models

Classification Method	Sensitivity (%)	Specificity (%)
DT	0	50
LR	0	50
LDA	50	50
SVM	100	66.7
kNN	100	66.7
RF	100	66.7

wrong prediction of one (01) COVID-19 infected ultrasound video sample caused the SVM, kNN and RF classifiers to achieve their overall accuracy only till 75%. Nevertheless, these three (03) classifiers are highly sensitive towards the parameter settings that may cause them to have better training and testing accuracies compared to the DT, LR and LDA classification models. The testing strength and performance of each selected classification model can be examined by calculating the sensitivity and specificity using:

$$Sensitivity = TP/(TP + FN) \tag{4.7}$$

$$Specificity = TN/(TN + FP) \tag{4.8}$$

where, *TP* is True Positive values, *TN* is True Negative value, *FP* is False Positive value, and *FN* is False Negative value. The calculated sensitivity and specificity values of each classification model are given in Table 4.1.

According to the calculated sensitivity and specificity values as shown in Table 4.1, the DT and LR classifiers have the lowest sensitivity. Thus, these two (02) classifiers have the 100% performance in predicting the normal subject lungs, whilst both COVID-19 infected and normal subject's prediction rate of the LDA classifier remains the same. The results of Table 4.1 are further illustrated in Figure 4.7.

Nevertheless, the more important task of this study is to predict the COVID-19 infected lungs. Therefore, all six (06) classifiers can be specified into two (02) categories by comparing their sensitivity and specificity values. The category one (01) includes the classifiers which have best sensitivity and specificity (i.e. SVM, kNN and RF), whilst category two (02) comprise the classifiers which have the least sensitivity and specificity (i.e. DT, LR and LDA). The final prediction results of the test ultrasound video samples of category one (01) classifiers and category two (02) classifiers are shown in Table 4.2 to evaluate their performance further.

The results of Table 4.2 show that the COVID-19 (2) test ultrasound video sample always carries the wrong prediction (predict COVID-19 (2) patient as a

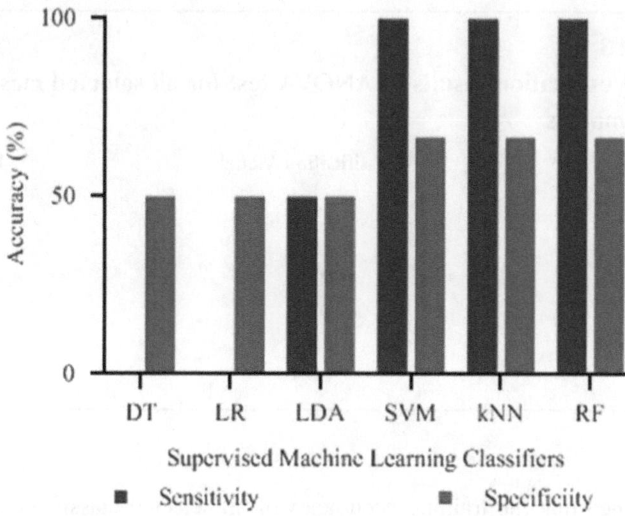

FIGURE 4.7 Sensitivity and Specificity of selected classifiers: DT, LR, LDA, SVM, kNN and RF.

TABLE 4.2
Prediction results of classification models (C – correct, W – wrong)

Category	Ultrasound Video Sample	COVID-19 (1)	COVID-19 (2)	Normal (1)	Normal (2)
One (01)	SVM	C	W	C	C
	kNN	C	W	C	C
	RF	C	W	C	C
Two (02)	DT	W	W	C	C
	LR	W	W	C	C
	LDA	C	W	C	W

normal subject) in all three (03) classifiers in category one (01). Similarly, the same COVID-19 (2) sample has the wrong prediction (predict COVID-19 (2) patient as a normal subject) results in all three (03) classifiers of category two (02). From these results, an assumption can be made which says the pattern of the region-wise edge factor-based motion vectors of the COVID-19 (2) sample may be different from the training samples. The prediction results of Table 4.2 show that the DT and LR classifiers wrongly predict both COVID-19 (1) and COVID-19 (2) ultrasound video samples (predict both COVID-19 samples as normal samples), whilst the LDA classifier correctly predicts COVID-19 (1) sample and wrongly classifies Normal (2) sample (classifies Normal (2) samples as a COVID-19 sample). Subsequently, it

TABLE 4.3
P-value estimation results of ANOVA test for all selected classification models

Category	Classification Model	P-value
One (01)	SVM	0.478
	kNN	0.478
	RF	0.478
Two (02)	DT	0.666
	LR	0.666
	LDA	0.666

can be assumed that the training accuracies of all selected classifiers may have a high impact on these false prediction results. The training accuracies of all three (03) classification models in category one (01) have 100% accuracy, and the DT, LR and LDA models have 50%, 33.3% and 50% training accuracies, respectively. It can be noticed that the training accuracies of classification models in category two (02) (i.e. DT, LR and LDA) are lower than category one (01) classification models (i.e. SVM, kNN and RF). The category one (01) classification models provide a 75% test accuracy for COVID-19 and normal subject's lung ultrasounds videos, whilst category two (02) classification models provide 50% test accuracy. These results show that category one (01) classification models can successfully perform training and testing processes. However, the selected database carries ten (10) ultrasound videos that have appropriate visible patterns to investigate both COVID-19 and normal subjects' lung conditions. Therefore, all these possible ultrasound videos have given acceptable results for this study.

The obtained classification results of Table 4.2 are further validated using a single factor ANOVA test to evaluate the performance of selected six (06) classification models through a hypothesis test. The ANOVA test is well-known as a variance analysis test, which provides information on the relationship between a test datasets' dependent and independent variables. Therefore, the significant differences between the variables are observed by calculating the P-value (probability value) by considering the significant level as 0.5. The P-value estimations are provided in Table 4.3 for all selected classification models in Table 4.2.

The P-value estimation results, shown in Table 4.3, clearly provide the information of acceptance and rejection of null hypothesis (H_0). Thus, it further proves that the classification models in category one (01) have the potential to determine the best classification results from the input feature selections. This is proved by the P-values estimation of 0.478 for all three (03) classification models (i.e. SVM, kNN, and RF), which is less than the significant level of 0.5. Eventually, these classification models reject the null hypothesis, whilst all three (03) classification models in category two (02) (i.e. DT, LR, and LDA) accept null-hypothesis due to their P-value estimation of 0.666, which is greater than the significance level of 0.5. The null-hypothesis test is a commonly applied method that helps in statistical inference. Therefore, this hypothesis test is used to evaluate the defined concepts, explanatory principles and results verification. Consequently, the statistical test evaluation results in Table 4.3 show that the category one (01) classifiers provide statistically significant results in classifying COVID-19 and normal subjects' lung ultrasound videos.

4.4 CONCLUSION

Many patients are not diagnosed with respiratory diseases at an early stage and are unaware of theirs carrying a disease. Due to their lack of awareness, any mild lung disease may cause severe damages to the lungs and respiratory system. The early prevention of respiratory diseases is needed to decrease their vilest versions and reduce the risk to life. The proposed methodology of this study can predict the COVID-19 infected patients by lung ultrasound video analysis. Prior to the feature analysis, the videos are sent through NRB algorithm to convert them as image frames and to extract features. In the NRB algorithm, the changes in the pixels of the extracted consequent image frames are examined by Adaptive Rood Pattern Search (ARPS) algorithm. Furthermore, the region-wise edge factor-based motion vectors are extracted from all the frames. The results achieved by the region-wise analysis of extracted motion vectors show that the fourth (04) and seventh (07) regions have dynamic variations compared to other regions, which means that COVID-19 virus may affect the functionality of all the regions except regions four (04) and seven (07). These results are further analyzed using Moving Exponential Growth (MEG) statistical function and found that the motion vector analysis has a high impact on identifying the functional patterns of COVID-19 infected lungs from the normal lungs.

Moreover, the computed MEG values are examined by applying DT, LR, LDA, SVM, kNN and RF supervised classifiers and achieved acceptable classification accuracies only from SVM, kNN and RF classification models. Results obtained from the DT, LR and LDA classification models are not exactly accurate as they provided a very low sensitivity and specificity values, especially both DT and LR models have zero (0) sensitivity rate. However, the sensitivity of SVM, kNN and RF classification models are extremely high. It can be seen from the results, the performance of the SVM, kNN and RF classification models are identical. Thus, it can be concluded why the NRB algorithm has caused an enormous influence to select the best features for the classifiers.

The achieved classification results can be used to warn COVID-19 infected people early on to take the proper precautions to prevent theirs being in danger of a severe lung problem. Additionally, this method is fast and convenient to be used by any user to predict their lung behaviour. Given ultrasound is a widely used cost-effective method in clinical studies to diagnose various diseases, including lung damages, however, due to each ultrasound video length, the time span of the clarification method proposed in this study may vary. Subsequently, if an ultrasound video length is longer, more time will be taken to extract the frames and their features. Similarly, when the ultrasound video length is shorter, less time will be taken to perform all feature extraction and classification tasks. Nevertheless, this proposed method shall be improved further to achieve even better classification results in both real-time and offline ultrasound video analyzing applications in the healthcare system. Towards this purpose, the increase of train samples and test samples may be a well-suited solution to enhance the best and the most accurate results. Although this method does not explicitly predict a particular disease of the lungs, yet it can be used as a primary lung movement evaluation method. Having said that, this study provides an essential solution for its problem statement and by considering the proven results, it can be finally concluded that the motivation and objectives of this study are successfully fulfilled.

REFERENCES

[1]. M. Arabiat, A. E. Foderaro and A. T. Levinson, "Lung ultrasound for diagnosing patients with severe dyspnea and acute hypoxic respiratory failure," *Rh. I. Med. J. (2013)*, vol. 102, no. 10, pp. 34–38, 2019.

[2]. M. Encyclopedia, "Lung disease," pp. 1–3, 2020.

[3]. G. Manogaran, P. M. Shakeel, A. S. Hassanein, P. Malarvizhi Kumar and G. Chandra Babu, "Machine learning approach-based gamma distribution for brain tumor detection and data sample imbalance analysis," *IEEE Access*, vol. 7, no. c, pp. 12–19, 2019, 10.1109/ACCESS.2018.2878276.

[4]. M. G. M. Milani, P. E. Abas and L. C. De Silva, "Identification of normal and abnormal heart sounds by prominent peak analysis," *ACM International Conference Proceeding Series,* New York, USA, pp. 31–35, September 2019, 10.1145/3364908.3364924.

[5]. J. A. Cruz and D. S. Wishart, "Applications of machine learning in cancer prediction and prognosis," *Cancer Inform.*, vol. 2, pp. 59–77, 2006, 10.1177/117693510600200030.

[6]. J. Verma, M. Nath, P. Tripathi and K. K. Saini, "Analysis and identification of kidney stone using Kth nearest neighbour (KNN) and support vector machine (SVM) classification techniques," *Pattern Recognit. Image Anal.*, vol. 27, no. 3, pp. 574–580, 2017, 10.1134/S1054661817030294.

[7]. S. Roychowdhury, D. D. Koozekanani and K. K. Parhi, "DREAM: diabetic retinopathy analysis using machine learning," *IEEE J. Biomed. Heal. Informatics*, vol. 18, no. 5, pp. 1717–1728, 2014, 10.1109/JBHI.2013.2294635.

[8]. T. H. Wu, G. K. H. Pang and E. W. Y. Kwong, "Predicting systolic blood pressure using machine learning," *2014 7th Int. Conf. Inf. Autom. Sustain. Sharpening Futur. with Sustain. Technol. ICIAfS 2014*, vol. 2014, 2014, 10.1109/ICIAFS.2014.7069529.

[9]. G. D'Souza, H. H. Zhang, W. D. D'Souza, R. R. Meyer and M. L. Gillison, "Moderate predictive value of demographic and behavioral characteristics for a diagnosis of HPV16-positive and HPV16-negative head and neck cancer," *Oral Oncol.*, vol. 46, no. 2, pp. 100–104, 2010, 10.1016/j.oraloncology.2009.11.004.

[10]. R. Zhang, C. Cheng, X. Zhao and X. Li, "Multiscale mask R-CNN–based lung tumor detection using PET imaging," *Mol. Imaging*, vol. 18, pp. 1–8, 2019, 10.1177/1536012119863531.

[11]. J. M. Kong, "Cascade convolutional neural networks for automatic detection of thyroid nodules in ultrasound images,"*Am. Assoc. Phys. Med.*, vol. 45, no. 5, pp. 1678–1691, 2017, 10.1002/mp.12134.

[12]. A. M. Santos, A. O. De Carvalho Filho, A. C. Silva, A. C. De Paiva, R. A. Nunes and M. Gattass, "Automatic detection of small lung nodules in 3D CT data using Gaussian mixture models, Tsallis entropy and SVM," *Eng. Appl. Artif. Intell.*, vol. 36, pp. 27–39, 2014, 10.1016/j.engappai.2014.07.007.

[13]. B. S. Lin, "Using deep learning in ultrasound imaging of bicipital peritendinous effusion to grade inflammation severity," *IEEE J. Biomed. Heal. Informatics*, vol. 24, no. 4, pp. 1037–1045, 2020, 10.1109/JBHI.2020.2968815.

[14]. I. Banerjee, "Brain Tumor Image Segmentation and Classification using SVM, CLAHE AND ARKFCM" for Intelligent decision support systems, applications in signal processing, Chapter 3,978-3-11-062110-5, pp. 53-70, October 2019.

[15]. H. F. Routh, "Doppler ultrasound," *IEEE Eng. Med. Biol. Mag.*, vol. 15, no. 6, pp. 31–40, 1996, 10.1109/51.544510.

[16]. D. A. Lichtenstein, "Lung ultrasound," *Reanimation*, vol. 12, no. 1, pp. 19–29, 2003, 10.1016/S1624-0693(02)00005-1.

[17]. S. Liu , Y. Wang and X. Yang, "Deep learning in medical ultrasound analysis: a review," *Engineering*, vol. 5, no. 2, pp. 261–275, 2019, 10.1016/j.eng.2018.11.020.

[18]. M. Correa, "Automatic classification of pediatric pneumonia based on lung ultrasound pattern recognition," *PLoS One*, vol. 13, no. 12, pp. 1–13, 2018, 10.1371/journal.pone.0206410.

[19]. B. Zhou and X. Zhang, "Lung mass density analysis using deep neural network and lung ultrasound surface wave elastography," *Ultrasonics*, vol. 89, pp. 173–177, 2018, 10.1016/j.ultras.2018.05.011.

[20]. B. Zhou, B. J. Bartholmai, S. Kalra and X. Zhang, "Predicting lung mass density of patients with interstitial lung disease and healthy subjects using deep neural network and lung ultrasound surface wave elastography," *J. Mech. Behav. Biomed. Mater.*, vol. 104, no. 2, p. 103682, 2020, 10.1016/j.jmbbm.2020.103682.

[21]. M. Palacio, "Performance of an automatic quantitative ultrasound analysis of the fetal lung to predict fetal lung maturity," *Am. J. Obstet. Gynecol.*, vol. 207, no. 6, pp. 504.e1–504.e5, 2012, 10.1016/j.ajog.2012.09.027.

[22]. K. N. Bhanu Prakash, A. G. Ramakrishnan, S. Suresh and T. W. P. Chow, "Fetal lung maturity analysis using ultrasound image features," *IEEE Trans. Inf. Technol. Biomed.*, vol. 6, no. 1, pp. 38–45, 2002, 10.32388/jwiq19.

[23]. R. La Torre, E. Cosmi, M. H. Anceschi, J. J. Piazze, M. D. Piga and E. V. Cosmi, "Preliminary report on a new and noninvasive method for the assessment of fetal lung maturity," *J. Perinat. Med.*, vol. 31, no. 5, pp. 431–434, 2003, 10.1515/JPM.2003.067.

[24]. E. Bonet-Carne, "Quantitative ultrasound texture analysis of fetal lungs to predict neonatal respiratory morbidity," *Ultrasound Obstet. Gynecol.*, vol. 45, no. 4, pp. 427–433, 2015, 10.1002/uog.13441.

[25]. Y. Du, Z. Fang and B. D. J. Jiao, "Application of ultrasound-based radiomics technology in fetal lung texture analysis in pregnancies complicated by gestational diabetes or pre-eclampsia," *Ultrasound Obstet. Gynecol.*, vol. 57, no. 5, pp. 804–812, 2020, 10.1002/uog.22037.

[26]. C. Mehanian and K. M. Wilson, "Deep learning-based pneumothorax detection in ultrasound videos," in *Lecture Notes in Computer Science*, Springer, China, 2019.

[27]. P. M. Cheng and H. S. Malhi, "Transfer learning with convolutional neural networks for classification of abdominal ultrasound images," *J. Digit. Imaging*, vol. 30, no. 2, pp. 234–243, 2017, 10.1007/s10278-016-9929-2.

[28]. S. Kulhare, X. Zheng, C. Mehanian, C. Gregory, H. Xie, J. M. Jones and B. Wilson, "Ultrasound-based detection of lung abnormalities using single shot detection convolutional neural networks," in *Lecture Notes in Computer Science*, Springer, Cham., 2018.

[29]. J. C. M. Than, "Lung disease stratification using amalgamation of Riesz and Gabor transforms in machine learning framework," *Comput. Biol. Med.*, vol. 89, no. 6, pp. 197–211, 2017, 10.1016/j.compbiomed.2017.08.014.
[30]. L. Valdes-Burgos, S. L. Contreras-Ojeda, J. A. Domínguez-Jiménez and J. Lopez-Bueno, "Analysis and classification of lung tissue in ultrasound images for pneumonia detection," 2020, 10.1117/12.2542615.
[31]. M. P. Prasanth, G. Rashmi and B. R. Srilekha, "Eye movement detection for paralyzed patient using pressure sensor," *Int. J. Eng. Res.*, vol. 7, no. 11, pp. 395–400, 2016.
[32]. S. K. Veeramani and E. Muthusamy, "Detection of abnormalities in ultrasound lung image using multi-level RVM classification," *J. Matern. Neonatal Med.*, vol. 29, no. 11, pp. 1844–1852, 2016, 10.3109/14767058.2015.1064888.
[33]. A. Roopashree, "Machine learning approach: Enriching the knowledge of Ayurveda from Indian Medicinal Herbs in the book Challenges and applications of Data analytics in social perspectives (ISBN:290719-084254), November 2020.
[34]. A. Mckeever, *Here's What Coronavirus does to the Body*. National Geographic, 2020.
[35]. *WHO Coronavirus Disease (COVID-19) Dashboard*. World Health Organization, India, 2020.
[36]. M. R. Suma, "Hybrid cloud-Intra domain data security and to address the issues of interoperability," *Int. J. Recent Technol. Eng.*, vol 8, no. 5, pp. 340–344
[37]. C. P. Loizou and C. S. Pattichis, "Diffusion Despeckle Filtering," 2008.
[38]. R. Dass, "Speckle noise reduction of ultrasound images using BFO cascaded with Wiener filter and discrete wavelet transform in homomorphic region," *Procedia Comput. Sci.*, vol. 132, pp. 1543–1551, 2018, 10.1016/j.procs.2018.05.118.
[39]. H. A. Choudhury and M. Saikia, "Block matching algorithms for motion estimation: a performance-based study," *Lect. Notes Electr. Eng.*, vol. 347, pp. 149–160, 2015, 10.1007/978-81-322-2464-8_12.
[40]. Y. Nie and K. K. Ma, "Adaptive rood pattern search for fast block-matching motion estimation," *IEEE Trans. Image Process.*, vol. 11, no. 12, pp. 1442–1449, 2002, 10.1109/TIP.2002.806251.
[41]. A. V. Paramkusam and V. Arun, "A survey on block matching algorithms for video coding," *Int. J. Electr. Comput. Eng.*, vol. 7, no. 1, pp. 216–224, 2017, 10.11591/ijece.v7i1.pp216-224.
[42]. M. M. Umar and L. C. De Silva, "Fire boundary detection method using a unique structure from motion for Non-Rigid Bodies Algorithm (SFM-NRBA)," *ACM Int. Conf. Proceeding Ser.*, pp. 74–78, 2017, 10.1145/3163080.3163100.
[43]. M. M. Umar and L. C. De Silva, "Onset fire detection in video sequences using region based structure from motion for nonrigid bodies algorithm," *IET Conf. Publ.*, vol. 2018, no. CP750, pp. 1–4, 2018, 10.1049/cp.2018.1533.
[44]. "Exponential Growth and Decay," *MathBitsNotebook.com*.
[45]. M. Partridge and M. Jabri, "Robust principal component analysis," *Neural Networks Signal Process - Proc. IEEE Work.*, vol. 1, no. 5, pp. 289–298, 2000, 10.1201/b20190-2.
[46]. R. Zou, M. Schonlau and D. Ph, "Applications of Random Forest Algorithm Stata Syntax Classification Example: Credit Card Default Regression Example: Consumer Finance Survey," pp. 1–33.
[47]. H. H. Zhang, "Lecture 6: Binary Classification (III) – LDA and Logistic Regression Two Popular Linear Models for Classification," no. Iii, 2013.
[48]. H. Li, F. Qi and S. Wang, "A comparison of model selection methods for multi-class support vector machines," *Lect. Notes Comput. Sci.*, vol. 3483, no. IV, pp. 1140–1148, 2005, 10.1007/11424925_119.
[49]. P. Thanh Noi and M. Kappas, "Comparison of random forest, k-nearest neighbor, and support vector machine classifiers for land cover classification using sentinel-2 imagery," *Sensors (Basel)*, vol. 18, no. 1, pp. 1–20, 2017, 10.3390/s18010018.

5 Enhanced Effective Generative Adversarial Networks Based LRSD and SP Learned Dictionaries with Amplifying CS

K. Elaiyaraja[1], *Dr. M. Senthil Kumar*[2], *and B. Chidhambararajan*[3]

[1]Assistant Professor, Department of Information Technology, SRM Valliammai Engineering College, Kattankulathur

[2]Associate Professor, Department of Computer Science and Engineering, SRM Valliammai Engineering College, Kattankulathur

[3]Principal, SRM Valliammai Engineering College, Kattankulathur

CONTENTS

DOI: 10.1201/9781003194415-5

5.1 INTRODUCTION

In the era of medical tools and advanced diagnostic technology, image fusion is a primary and dominant tool. The various wide-ranging modalities like MRI, PET and CT are used to obtain images. Each modality has their own pros and cons with different purpose in order to scan objects. The information retrieved from a single modality image is not sufficient because of its intended use, affected by the inadequate spatial resolution and parameters like different scanning time. So, this insufficient image information leads to inaccurate detection of disease symptoms; in other words, undersized anatomical landmark segmentation and pathologies. Over many decades, numerous methods and techniques have been invented to get better resolution of source images [1]. The traditional methods like cubic interpolations, and its versions, work with an original image which data has to be preserved causing sharp edge information of boundaries to fail to achieve better resolution as many edges have a gradual. The SR (Super Resolution) reconstruction methodologies were introduced and became popular in the field of resolution enhancement where the original data might not even exist. The better PSNR (Peak Signal-to-Noise Ratio) was done [2] with the single image SR technique, which was proposed by using Random-forest model collection approach [3] and the high resolution-from-low resolution images were obtained. The Multi-Kernel vector regression-based SR technique was proposed specifically for the decomposition of source images. The experimental results from the SR technique were still not able to bring back the eminence images even though these techniques were considered as effective.

Image fusion in the medical era of technical tools and advanced technologies has become an effective information merging method and numerous techniques were proposed in the recent years. The MTA (Multi-scale Analysis) is a popular and commonly used method which includes DWT, ST, Curvelet and Contourlet transform (CT) and NSCT (Non Sub-sampled CT) [4]. To decompose a source image is to convert it into various sub-bands. Then, these divided-by-sub-bands-components are fused together based on the fusion technology used. But the entirety of pixels or coefficients are not always found and leads to loss of significant information resulting in poor image quality.

Due to superior advancements in the DL arena in medical image processing, some of the new SR based approaches and Sparse Representation (SpR) were proposed. The SpR based technique discovers the sparse regions from the source images using the given dictionary. The target image is obtained by merging the coefficients and the dictionary [5]. But in these methods, usually one and only dictionary is used to signify various morphological areas of the input images. The suboptimal illustration of intrinsic details is the result of these techniques. Even though these methods are successfully implemented, still the target images are obtained with noise. In order to avoid this issue, de-noising the source image is the initial strategy followed by fusion. By implementing this method, the artifacts created from de-noising propagate and magnify while performing fusion [6]. This chronological method may cause information loss.

The methods to conduct de-noising and image fusion at the same time might produce good results. The concurrent methods proposed for suppressing noise as well as retaining significant information were not always dovetailed to give results [7]. A precise SR method was proposed by CNN based on VGG-net. CNN (Convolutional Neural Networks) was introduced first for a single-image SR and invented SRCNN technique. For grayscale based medical images, the fundamental composition of SR and CNN methods was proposed in [8].

The Residual NN (ResNet) was introduced for image fusion and 52 layer-recursive networks were proposed [4]. In this approach, SR performance was improved by Residual NN. The unnecessary models were detached or removed while enlarging model size in Residual NN and were shown to have reasonable improvements. The Residual-dense block [9] and additionally taken up with results to train much deeper networks along with channel consideration. The PSNR performance was successfully achieved in this method.

SR based medical images are still considered as an open challenge even though various methods or technologies have been proposed in recent decades. The reconstructions of medical images are still not achieved for high up-scaling subject. In this proposed method, the image is modeled with low-rank sparse technology and it removes noise from the ruined images based on LRSD and DL [10]. To generate LR and Sparse dictionaries, alongwith influential classification capability, enforcing LR and sparse restrictions over the LR sparse mechanisms is needed. The learned dictionaries are formulated for sparse-coding model. Notably, noises are suppressed from the source images by improving weighted-nuclear standard on the part of the sparse model. The source images are decomposed into low rank and sparse modules, followed by the coefficient representations using its corresponding dictionaries being obtained. While fusing images, the coefficients of the similar modules from different source images are merged by the fusion rule called "max-absolute". Next, the fused or merged LR sparse modules are reconstructed by integrating the coefficients and its corresponding dictionaries [11]. The SR technique using GAN framework is initiated. The CS block (Compression and Stimulation Block) is increased by gaining significant information during decreasing unwanted information. Embedding CSB with EDSR is carried out. Finally, a novel fusion loss gains the limitations on low features to obtain the SR model.

The proposed EEGAN based SR methods is applied on four medical image datasets using increased CS Block, with the fusion loss and results from comparison with other existing models visualized.

5.2 RELATED WORK

In SpR and LR learning, choosing a dictionary D is mandatory and decisive as the functionalities of these kinds of learning methods are highly dependent. Usually, D can neither be logically intended nor learned from a training data [12]. The curvelets are chosen as a dictionary to find or discover the smooth arcs, boundaries and anisotropic compositions. The DCT dictionary is selected to explore the textural content as it can signify intermittent structures like texture sparsely [13]. The logical dictionaries are generally not adaptable and not sufficient to categorize natural image compositions.

A dictionary learned from training data patches is a standard approach for image restoration and k-means clustering. For pattern classification and recognition, the DL is also applied [14]. To implement relationship among various, either heterogeneous or homogeneous information sources, the multimodal-task driven DL technique is applied. A Bayesian approach for learning DD (Discriminative dictionary) was introduced in order to build a relationship between the atoms as well as class labels for dictionary. The DSC (Discriminative Sparse Coding) technique was proposed for increased discriminability with added robust classification. LR DDL (Low rank double DL) model was proposed in which one learns a discriminative D from ruined data.

5.2.1 SpR and DL

A secondary dictionary constructing model is implemented for every cluster. An entire dictionary is created by combining the training outputs once the secondary dictionary is learnt. An image fusion model using GP representation (Group-Sparse) is implemented, and DL constructed. A compact DL method is presented for multi modal based image fusion. To get the fundamental characteristics of source images and retain the hierarchical compositions of inactive wavelet, the sub-bands are proposed [12]. This method is not a robust one as it is based on MTA and represents various components of input images through only one dictionary. Images generally require structures with various spatial morphologies. Huge dictionaries and extra added training data are required in order to restructure an improved quality image.

5.2.2 The Producer and the Discriminator

5.2.2.1 Discriminative LR and Sparse DL

The reconstruction or restructure must be assured first in DL.

Let x_t € Y^n (t = 1,2,3….m) be sample or training data composed by reorganizing the tth image territory of training data x into a column-based vector [15]. In order to decompose the input image in to LR and Sparse modules, dictionaries along with restructuring capability can be learned from the following:

$$a = \begin{pmatrix} LR_D, & SP_D \\ LR_C, & SP_C \end{pmatrix} \tag{5.1}$$

$$a = AM_a \left\{ \sum_{t=1}^{N} \left(\| s_i - LR_D LR_{C,t} - SP_D LR_{C,t} \| \frac{2}{2} \right) + BP_1 \| LR_{C,t} \|_1 + BP_2 \| LR_{C,t} \|_1 \right\} \tag{5.2}$$

Where LR_D € T^M represent the LR and SP dictionaries. LR_C € T^{MxN} & SP_C € T^{MxN} denote the sparse coefficient metrics. BP_1 and BP_2 are referred as parameter balance. In LR and SP decomposition technique, functionality is reliant on the dictionaries LR_D and SP_D. Actually, the learned dictionaries can discriminate the LR and SP structures from the source image [16]. The following optimization can obtain the above.

$$a = AM_a\{\|s - LR_D LR_C - SP_D SP_C\|_G^2 + BP_3\|LR_D LR_C\|*\}$$
$$+ \{BP_4\|SP_D SP_C\|_1 + BP_1\|LR_C\|_1 + BP_2\|LR_C\|_1\} \tag{5.3}$$

For LR module $LR_D SP_C$ of a particular image, all feature columns in $LR_D SP_C$ should be correlated and be located in a subspace of least dimension. The $\| LR_D SP_C \|*$ is applied on the optimized model. LR_D is learned to explore the LR modules of a particular source image and its low rank. The restructured SP modules should be sparse and apply low rank restriction on LR_D and the regularized sparse on $SP_D SP_C$ in order to make this dictionary compact as well as favorable to partition the LR from SP modules. The derived objective functions are as follows.

$$S = [s_1, s_2, s_3, \ldots. s_N]$$
$$a = AM_a\left\{\|S - LR_D LR_C - SP_D SP_C\|_G^2 + BP_3\|LR_D LR_C\|_* + BP_4\|SP_D SP_C\|_1\right.$$
$$\left.+ BP_5\|LR_D\|_* + BP_1\|LR_C\|_1 + BP_2\|SP_C\|_1\right\} \tag{5.4}$$

From the above stated function, the learned dictionaries are discriminative for partitioning the LR from SP modules.

5.3 PROPOSED WORK

5.3.1 Decomposing LR and SP

The $LR_D SP_D$ learned dictionary can partition the input images into corresponding LR and SP modules. For the fused target image without any noises, the following decomposition technique can be used:

$$b = (\widehat{SP_C}, \widehat{LR_C})$$
$$b = AM_b\{\|R - \widehat{LR_D} - SP_D\widehat{SP_C}\|_G^2 + P_3\|LR_D\widehat{LR_C}\|_* + P_3\|LR_D\widehat{LR_C}\|_* + P_1\|\widehat{LR_C}\|_1$$
$$+ P_2\|\widehat{SP_C}\|_1\} \tag{5.5}$$

Here, the P_1, P_2 and P_3 are referred as scalar parameters. $R= [r_1, r_2, \ldots r_t]$ € T^{MxN} and r_t is considered as rth vector formed by reconstructing the rth patch of the input image. $LR_C SP_C$ are referred as corresponding coefficients.

If the source image is ruined by noise, then the noise will be merged with the SP modules as the modules are random as well as linearly uncorrelated. The NFM LR matrix can remove noise from this corruption. The reconstructed and accurate results are highly nonstandard as NNM indulgence work on various rank modules equally and lead to shortage of information that is information loss. The WNN (Weighted Nuclear Norm) assigns various weights to various solitary values and can solve this issue. For noise decomposition, the WNN technique is used as it is considered to be robust to the noise. The redefined form is shown below based on WNN:

$$b = AM_b\{\|Y = LR_D\widehat{LR_C} - SP_D\widehat{SP_C}\|_G^2 + P_3\|LR_D\widehat{LR_C}\|_* + P_4\|SP_D\widehat{SP_C}\|_{U,*} + P_1 \|\widehat{LR_C}\|_1 + P_2\|\widehat{SP_C}\|_1\} \tag{5.6}$$

Here, the WNN is defined as ||.||u,*=∑i|σ$_i$ω$_i$(.)| and σ_i is considered as the ith remarkable data of '.' and ω_i which is greater than zero represents the ith weight to σ_i. The parameter P_4 controls the involvement in the WNN. In order to retain the edge details of input image during de-noising, the WNN assigns various weights to various particular or remarkable data. While removing noise, significant information may smoothen out. The boundary details of an image are nothing but sparse modules and fusion techniques can enhance this significant information. The boundary details are retained and can be enhanced with the help of following method.

$$(b) = AM_B\{\|R - LR_D\widehat{LR_C} - SP_D\widehat{SP_C}\|_G^2 + P_4\|SP_D\widehat{SP_C}\|_{W,*} + P_3\|LR_D\widehat{LR_C}\|_* + P_5 \|SP_D\widehat{SP_C}\|_1 + P_1\|\widehat{LR_C}\|_1 + P_2\|\widehat{SP_C}\|_1\} \tag{5.7}$$

The primary function of the CS block is to adapt and recalibrate the features of source images by forming interdependencies among channels. Initially, the CS block compressing feature as follows

$$E_d = \frac{1}{IV}\sum_{t=1}^{I}\sum_{u=1}^{V} y_d(t, u) \tag{5.8}$$

Where, E_d denotes the compressed feature corresponding to the dth feature map y_d. 'I' denotes height and 'V' denotes the width of y_d. The compressed features are given as input to a three layered NN. The output dimension is same as dimension of the layer. The following expression depicts the actual CS block function:

$$V_s = O(E) = \sigma(V_2\delta(V_{1,E})) \tag{5.9}$$

Where $V_s = [v_1, v_2, v_1, \ldots v_s]$ represents the scale vector of actual feature maps and O (.) denotes the actual activation function. E= $[e_1, e_2, e_1 \ldots e_s]$ represents the feature vector of source image. σ denotes the sigmoid function and ʒ represents the ReLU function. V_1 and V_2 represent the weight of the source layer and destination layer, respectively. The final outcome of CS block is attained with:

$$\overline{y_d} = y_d \cdot V_d \tag{5.10}$$

Here the element-wise layers are represented by '.', or the dot. The actual activation function doesn't exploit the response of the concealed layer. O (E) starts its series from zero to one. The several CS blocks are fixed in a NN, and O (E) of (0, 1) will construct the response of the center layers. This will corrupt the act of the network. In order to avoid this disadvantage, the following function is used to get the improved CS block.

$$s = O_{im}(E) = \{l_1 \times [\sigma(V_2\delta(V_1, E))] + l_2 \times \sigma(E)\} \times 2 \tag{5.11}$$

Notable improvements in CS block:

- The enduring function of (4) makes use of inputs and the outputs of the three-layered network. The complexity faced in the training routine is reduced.
- The weakening of significant information caused by performing several manipulations can be reduced.

5.3.2 Enhanced Effective Generative Adversarial Networks

The SR technique comprises GAN framework and the boosted CS block. This block is attached with producer and divider, followed by the fusion function. The Figure 5.1 depicts the proposed method.

5.3.3 The Producer and the Discriminator

The EDSR technique implemented by Lim et al. was simplified to perform producer of GAN model. The novel EDSR has Resblocks and kernels. Then, the boosted CS block was appended with the CNN. The discriminator comprises eight convolutional layers with kernels. Later, the fusion block merges the extracted features from the least 3 layers. The discriminator processes the low level frequency details, and the unwanted details are subsequently reduced. The classification process is accomplished by processing pooling function, convolution constructions and Sigmoid triggering function. The following block diagram Figure 5.2 depicts the proposed method.

5.3.4 Fusion Scheme

The WNN constriction in the image decomposition technique, the learned SP_C, is clear. The reformed image done by SP_CSP_D is noiseless. LR_CSP_C are the coefficients with higher norm values which correspond to the patches with prominent information.

The fusion operation for training GAN technique is performed. The loss function is used to combine loss (L), comparative opposition loss, sensitivity loss and MSEL (Mean Square Error loss).

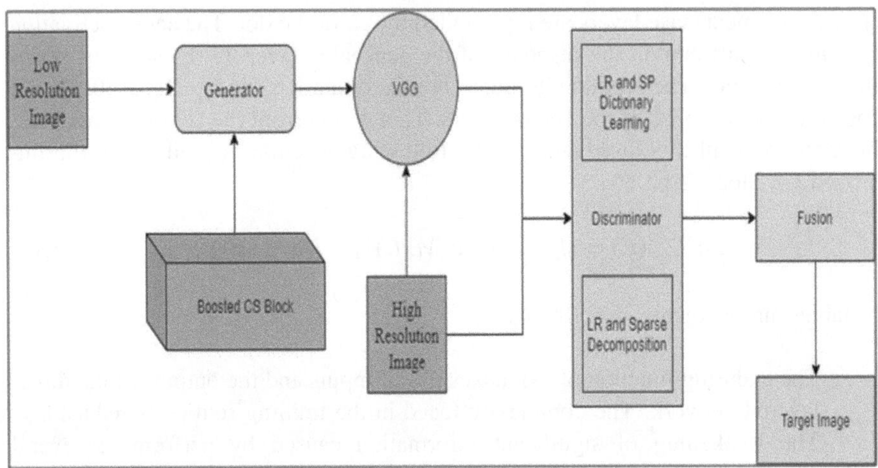

FIGURE 5.1 Architecture Diagram of Proposed Method.

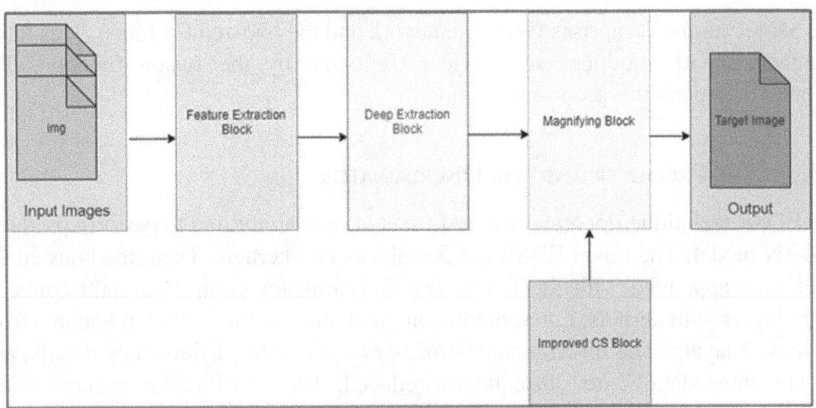

FIGURE 5.2 Block Diagram of Proposed Method.

$$T_{img} = C_L + C_L CO_L + S_L + C_{ME} ME_L \tag{5.12}$$

Here, T_{img} represents the target image, and C, C_L, C_{ME} are constants. These constants are used to represent hyper parameters which control the loss. The comparative opposition loss adds to top-level semantic features instead of pixel-level features. This feature detail is same as the perceptual resemblance. The variable C_L implements the network to retrieve details from the base source images. The other two variables ($C_L CO_L + S_L$) lead to increased PSNR for the reconstructing image. The last variable ($C_{ME} ME_L$) is used to reduce the MSE between the targeted image and its corresponding source image. The algorithm used for Discriminative LR Sparse DL, and Low Rank and Sparse Decomposition are given in the Figures 5.3 and 5.4, respectively.

Algorithm for Discriminative LR Sparse DL

Step 1: LR_D SP_D and BP (t=1 to Maximum iteration)

Step 2: Initialize the Balance parameters

Step3: Update Z_l =AM {|| Z-Z_l||+||Z_l||*}|

Step 4: Update Z_l =AM {|| Z-Z_l||+||Z_l||1}

Step 5: Update LR_C =AM {|| Z- LR_D LR_C||+BF1 ||Z_l||1} and

Update LR_C =AM {|| Z- SP_D SP_C||+BF2 ||Z_l||1}

Step 6: Update X=AM((BF5/μ||z||*+1/2)X-(LR_D+V/μ||^2_F))

Step 7: LR_D=(J_l+2/ μ LR_C LR_C^M)$^{-1}$(2/ μX_l LR_C^M+X+U/ μ)

Step 8: Repeat step 3 if the iteration reaches maximum value

Step 9 : Output SP_D, LR_D

FIGURE 5.3 Algorithm for Discriminative LR Sparse DL.

Algorithm for Low Rank and Sparse Decomposition

Step 1: Initiate Variables P, Q and R
Parameters from 1…5
Step 2: X not satisfied then do
Step 3: Update Q_l and Q_s
Step 4: Update P=AM{|| Q-R||$_G^2$ + E||P||$_{w,*}$}
Step 5: LR_C=AM $_{LRC}${||P-$LR_D$$LR_C$||$_G^2$+E1|| LR_C||$_1$}
SP_C {||Q-$SP_D$$SP_C$||$_G^2$+E2|| SP_C||1}
Step 6: Repeat Step 3 until the iteration reaches its maximum value
Step 7: LR_C, SP_C

FIGURE 5.4 Algorithm for LR and Sparse Decomposition.

5.4 EXPERIMENTAL SETUP

The proposed Enhanced Effective Generative Adversarial Networks have been implemented using Ubuntu version 16.04 with PyTorch 0.4.1, NVIDIA 1080Ti, CUDA version 8 and CUDNN version 5.1. All experiments were done on a set of MRI datasets available from the internet. The 600 training dataset images were used to evaluate this proposed method. First, the effectiveness of the proposed method is processed by applying two algorithms. The Alg1 is applied for discriminating LRSP learning, and Alg2 is applied for LR and SP decompositions. Once the decomposition has been done, the producer and discriminator functions are applied followed by fusion technique.

Applying Alg1:

Fixing parameters: First, the number of iterations is fixed, and the parameters used for these iterations are (t = 1,2,3,4,5) followed by BP1 = 0.005, BP2 = 0.005 and B5 = 1 in the DL. To perform the evaluation on these training data, the parameter for iteration is set to 10. That is Alg 1 is repeatedly executed till it satisfies the condition Max <= 10. The BP3 and BP4 parameters are included to take control over the LR and SP constraints. The BP3 and BP4 had produced different parameters when the constant parameters were assigned to BP1, BP2 and BP3 parameters. The Figure 5.3 depicts the algorithm for Discriminative LR Sparse DL.

Applying Alg2:

After Alg1 has been processed, the Alg2 was applied and used up to 6 parameters and iterations. To obtain the value of noise in the image, the same procedure which is done in [10] was applied and, then, the P1, P2, P3 and P5 on the result image were examined. Initially, the value one is assigned to P3 and its post effect of P1 and P2 was examined. The resulting images during this process have been depicted in the Figure 5.5. If the P3 value was assigned to 1, satisfactory result from this second algorithm was obtained. Fixing P1 = 1, P2 = 1, P3 = 1 and P4 = 0, the result when P5 = 0.005, 0.01, 0.1, 1 are given in the Figure 5.5. If the number of iterations is set to 30 in this second algorithm, it produces noise-cum-grayscale images and five for spectral based images.

5.4.1 Applying Enhanced Effective Generative Adversarial Networks

The results produced by the above described second algorithm images are applied to this improved CS block. The parameters are determined as 0.8 and 0.2. As per the second algorithm, the constant value of C, C_L and C_{ME} are varying from zero to ten. The average PSNR and SSIM of this proposed technique on all these sample images, a scaling factor of four is used as shown in the following Table 5.1.

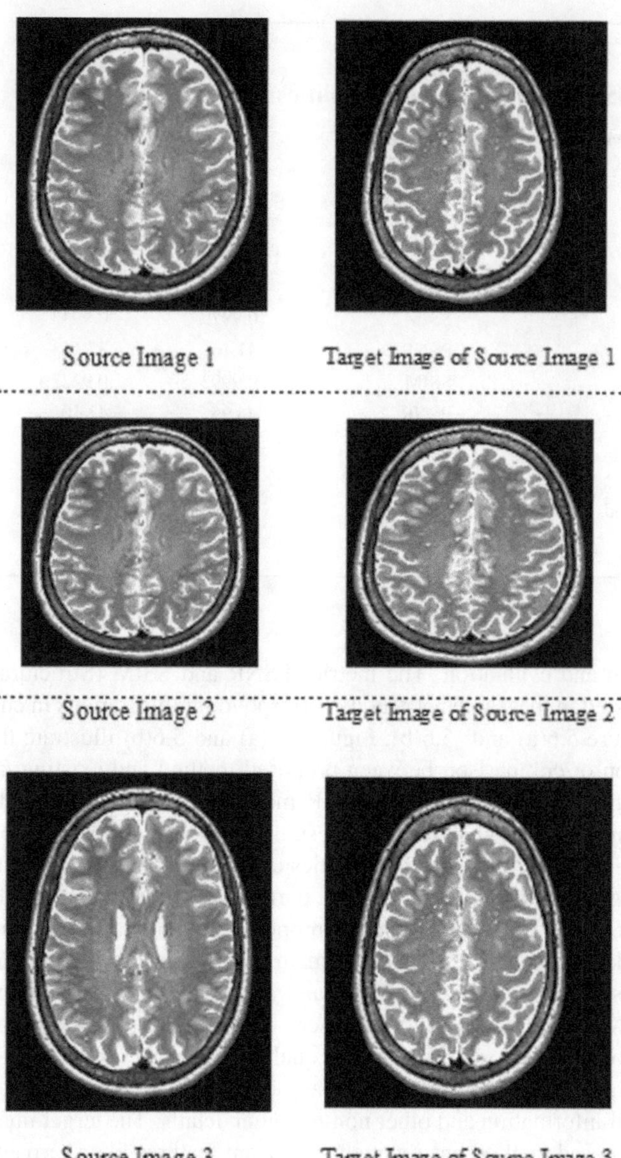

FIGURE 5.5 Source Image and its Corresponding Target Image.

5.5 DISCUSSION

The evaluation of the proposed method is done with the comparison of other existing models. The existing technique Bicubic and the State-of-the-art SR methods such as EDSR, SRGAN, DDBPN and VDSR were selected for comparison. Parameter wise, the same value is used for all the methods and these values are available in the base paper. Around 20 images from the training sample were opted

TABLE 5.1
PSNR and SSIM from different modalities

Technologies/Methods	Parameters	8	16	32
VDSR	SSIM	0.9975	0.9940	0.9937
	PSNR	37.92	33.43	32.23
SRGAN	SSIM	0.9501	0.9342	No result
	PSNR	26.43	26.3	No result
Bicubic	SSIM	0.9975	0.9943	0.9964
	PSNR	41.16	37.11	35.42
D-DBPN	SSIM	0.9993	0.9222	NA
	PSNR	44.95	42.10	NA
EDSR	SSIM	0.9993	0.9667	0.9741
	PSNR	44.97	29.91	29.64
Proposed method	SSIM	0.9994	0.9989	0.9986
	PSNR	43.98	28.34	26.36

for validation and evaluation. The metrics PSNR and SSIM (Structural Similarity Index) are listed in the Table 5.1 and its corresponding illustrations in chart form are given in Figure 5.6(a) and 5.6(b). Figure 5.6(a) and 5.6(b) illustrate the graphical representation of comparison between proposed method and existing methods.

When comparing the SSIM and PSNR metrics of proposed method with traditional system, our method shows good results. To further validate the outcome, the noise-laden images from various modalities and clear images obtained through this method are given in the Figure 5.5. The portability of this method is light-weight than the EDSR and DDBPN systems as proposed method has minimum number of layers than do the existing systems. The main advantage of this technique is that the simultaneous process of fusing images and de-noising can be accomplished. The effective GAN uses the LR and SP discrimination, and, LR and Sparse decomposition in order to improve the image quality. Then, the boosted CS block amplifies the significant features with the help of sparse dictionaries, and suppresses the unwanted information and other non-essential details. The target image obtained from the proposed method has retained significant features like sharp edges than its source image, and shows the best results. The best scores are highlighted in the Table 5.1 which indicates best performance.

5.6 CONCLUSION

In this proposed method, suppressing noises and significant data improvement using Enhanced Effective Generative Adversarial Networks have been done successfully. This technique is based on DL (Dictionary Learning) and LRSD (Low Rank Sparse-Decomposition). The input images are decomposed into LR and SP modules by using DL. The boosted CS blocks in the producer and discriminator are applied and

(a)

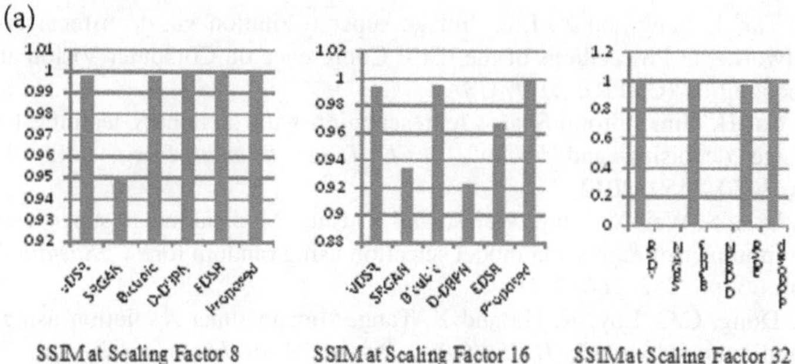

FIGURE 5.6A Comparison Chart of SSIM at Various Scaling Factors.

(b)

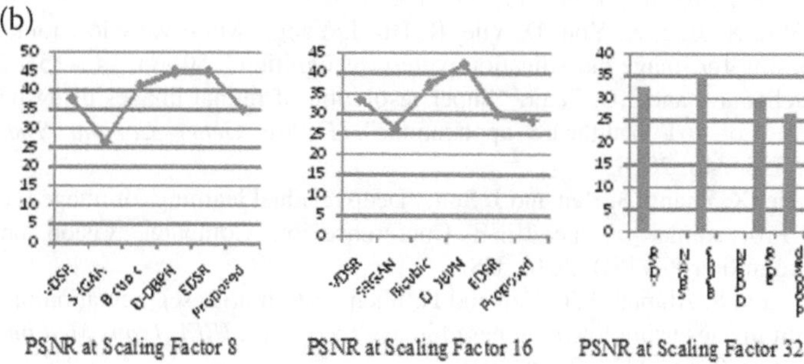

FIGURE 5.6B Comparison Chart of PSNR at Various Scaling Factors.

the results were shown for successful application. To attain maximum significant medical image details, the source image Compression and Stimulation (CS Block) block is developed by amplifying significant features. To fuse the source images and improve the target image features, the WNN and sparse limitation modules were used. The results obtained from the proposed method are apparently outperforming that of other existing methods. For future enhancement, the medical image restoration for blurred images can be used to implement with Super Resolution technique, alongwith spatial resolutions.

REFERENCES

[1]. I. Banerjee "Brain Tumor Image Segmentation and Classification using SVM, CLAHE AND ARKFCM" for Intelligent decision support systems, applications in signal processing, October 2019

[2]. J. Kim, J.K. Lee and K.M. Lee, "Accurate image super resolution using deep convolutional networks", in Proceedings of the IEEE Conference on Computer Vision and Pattern Recognition, 2016. USA.

[3]. T.M. Lehmann, C. Spitzer and K. Gonner, "Survey: Interpolation methods in medical image processing", *IEEE Trans Med Imaging*, vol. 18, no. 110, pp. 1049–1075, 1999.

[4]. Y. Tai, J. Yang and X. Liu, "Image super resolution via deep recursive residual network", in Proceedings of the IEEE Conference on Computer Vision and Pattern Recognition (CVPR), 2017. USA.
[5]. S. Lo, H. Yim "Group Sparse representation with dictionary learning for medical image denoising and fusion", *IEEE Trans Biomed Engg*, vol. 59, no. 12, pp. 3450–3459, 2012.
[6]. Q. Dou, S. Wei, X. Yang, W. Wu and K. Liu, "Medical image super resolution via minimum error regression model selection using random forest", *Sustain Cities Soc*, vol. 66, pp. 252–266, 2018.
[7]. D. Dong, C.C. Loy, K. He and X. Tang, "Image super resolution using dep convolutional networks", *IEEE Trans Pattern Anal Mach Intell*, vol. 38, no. 2, pp. 295–307, 2016.
[8]. D. Mahapatra, B. Bozorgtabar and R. Garnavi, "Image super resolution using progressive generative adversarial networks for medical image analysis", *Comput Med Imaging Graph*, vol. 71, pp. 30–39, 2019.
[9]. F. Wu, X. Jing, X. You, D. Yue, R. Hu, J. Yang, "Multi view low rank dictionary learning for image classification pattern recognition", 50, pp. 143–154, 2016.
[10]. J. Jebadurai and J.D. Peter, "Super resolution of retinal images using multi kernel SVR For IoT healthcare applications", *Future Gener Comput Syst*, vol. 83, pp. 338–346, 2018.
[11]. K. He, X. Zhang, S. Ren and J. Sun, "Deep residual learning for image recognition", in Proceedings of the IEEE Conference on Computer Vision and Pattern Recognition (CVPR), 2016. USA.
[12]. Z. Tang, S. Ahmad, P.T. Yap and D. Shen, "Multi atlas segmentation of MR tumor brain images using low rank based image recovery", *IEEE Trans Med Imaging*, vol. 37, no. 10, pp. 2224–2235, 2018.
[13]. H. Liu, J. Xu, Y. Wu, Q. Guo, B. Ibragimov and L. Xing, "Learning deconvolutional deep neural network for high resolution medical image reconstruction", *Inf Sci*, vol. 168, pp. 142–154, 2018.
[14]. F. Lin, J.D. Rojas and P.A. Dayton, "Super resolution contrast ultrasound imaging: Analysis of imaging resolution and application to imaging tumor angiogenesis", in Proceedings IEEE International Ultrasonics Symposium (IUS), 2016. USA.
[15]. S. Wei, X. Zhou, W. Wu, Q. Pu, Q. Wang and X. Yang. "Medical Image super-resolution by using multi dictionary and random forest", *Sustain Cities Soc*, vol. 37, pp. 358–370, 2018.
[16]. C. Zu, Z. Wang, D. Zhang, P. Liang, Y. Shi, D. Shen and G. Wu, "Robust multi atlas label propagation by deep sparse representation", *Pattern Recog*, vol. 63, pp. 511–517, 2017.

6 Deep Learning Based Parkinson's Disease Prediction System

Dr. G. Padmapriya[1], *Dr. R. Elakkiya*[2], *Dr. M. Prakash*[3], *and Vinoth Kumar M*[4]

[1]Associate Professor, Department of Computer Science and Engineering, Saveetha School of Engineering, SIMATS
[2]Assistant Professor, Center for Information Super Highway (CISH), School of Computing, SASTRA Deemed to be University
[3]Associate Professor, Department of Computer Science and Engineering, SRM Institute of Science and Technology
[4]Associate Professor, Department of Information Science and Engineering, Dayananda Sagar Academy of Technology & Management

CONTENTS

6.1 INTRODUCTION

Parkinson's disease is one among the chronic neurodegenerative (loss of structure or function of neurons) disorders which may result in continuous degeneration of functions that leads to serious disabilities in simple day to day activities in people,

DOI: 10.1201/9781003194415-6

due to defect in dopamine mechanism in brain cells. The neurons are the functional unit of the brain in human beings. A healthy and normal neuron is shown in Figure 6.1. Neuron has dendrites or axons, a cell body and a nucleus that includes DNA. If the neuron is sick, it will not have its extension with other neurons. Thus, it will not be able to communicate with other parts of the brain and its metabolism activity will be degraded [1,2].

a. **Factors Concerned In Parkinson's Disease:**

1. **Environmental Factors:**

   Depending on the location or surrounding in which the person lives that may affect the brain, the environmental factors are defined here. There is much evidence to prove that certain environmental factors that may lead to neurodegenerative disorders mainly Alzheimer's and Parkinson's are namely,

   i. Exposure to heavy metals like lead and aluminum
   ii. Exposure to pesticides
   iii. Air Pollution results in respiratory diseases
   iv. Biotic and Abiotic contaminants present in water directs to water pollution.
   v. Unhealthy lifestyle which leads to obesity and sedentary lifestyle.
   vi. Increase in psychological stress leads to increase the level of stress hormones which reduces the neurons' functionality.

2. **Brain Injuries or Biochemical Factors**:

   Brain is an overall control system for human beings. Certain people may be affected if they have injuries in the brain, so that a few biochemical enzymes are not secreted to neurons leading to the instability in human beings.

3. **Aging Factor:**

   Based on several recent researches in India, the main reason for the Parkinson's disease is aging.

4. **Genetic Factors:**

   Evidence suggests that the genetic factor is an important decision-maker

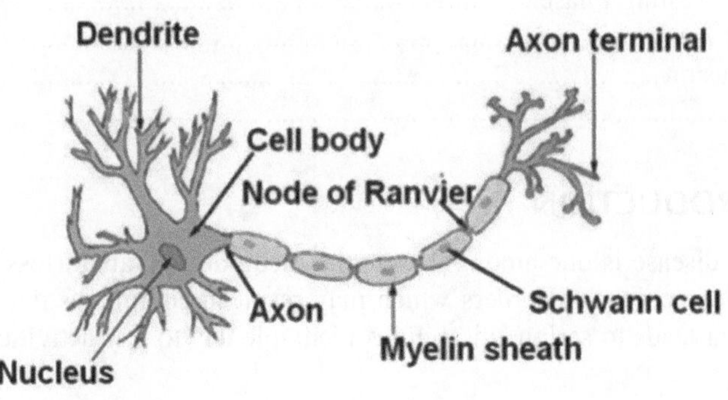

FIGURE 6.1 Structure of neuron in human brain.

of the disease; actions of diverse genes classify the level of neurodegenerative diseases and gradually increase overtime. Pharmacodynamics and pharmacokinetics are two categories of neurodegenerative diseases based on genetic factors.

5. **Smoking:**

S-E Soh et al. [3] proved that consuming tobacco in any form in people puts them at risk where they have 50 percentage chance of getting Parkinson's disease.

a. **The Parkinson Disease - Symptoms**

The Parkinson's disease shows symptoms as depicted in Figure 6.2 and categorized into:

1. Motor symptoms.
2. Non-motor symptoms.

    1. **Motor symptoms**

    J Jankovic [4] Tremor is the most common type of symptom for Parkinson's disease. Parkinson's gait affects the quality of life for the affected person. Changes in speech may be identified. Frozen feet will

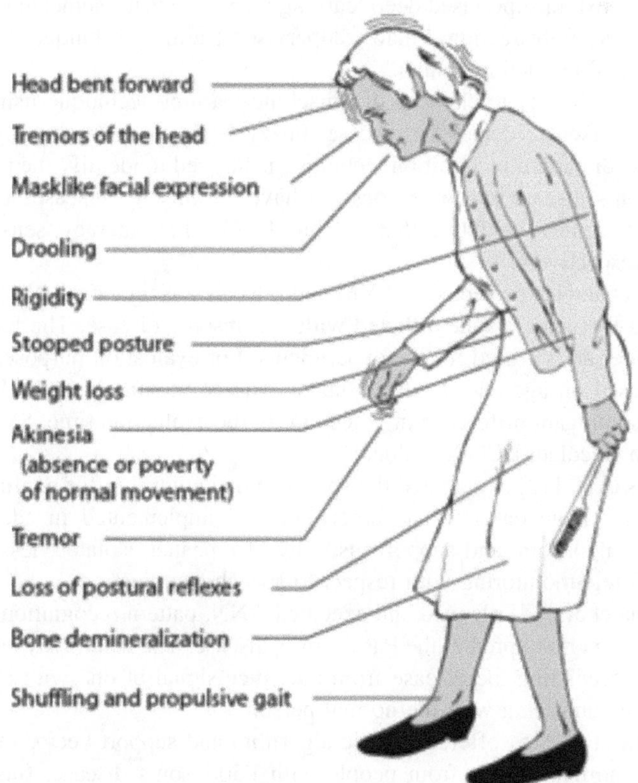

FIGURE 6.2 Symptoms of Parkinson's diseases.

make you feel that you cannot move your feet. There will be instability in body posture.

2. **Non-motor symptoms**

Non-motor symptoms include disorder in mood, depression, constipation, more sweating in body, cognitive changes, pain, problems in vision, change in body weight, change in sensing smell, sleep disorders [5].

6.2 LITERATURE SURVEY

Recently, there have been many advances in Parkinson's disease identification techniques to detect the disease at an early stage [6,7]. If Parkinson's disease is identified at an early stage, the impact and severity of the disease may be reduced. Based on voice data and speech signals, enormous amount of research has been made in pre-processing, feature selection, classification, with prediction techniques being implemented and developed using artificial intelligence techniques [8,9].

Mamoshina et al. [10] proposed a deep learning technique which involves multiple hidden layers to achieve a higher level of abstraction to predict the disease. Preprocessing is done first, in order to process the raw data and to achieve the quality data. Next, unsupervised deep learning technique is implemented to achieve the higher level of abstraction. Finally, supervised learning technique is applied to retrieve the final prediction results.

Indira R. et al. [11] implemented a machine learning technique using speech/voice of the person to detect the disease. Fuzzy C-means clustering algorithm in association with pattern recognition technique is utilized to identify the person who has Parkinson's disease and who does not have Parkinson's disease. The experimental results achieve 68.04%, 75.34%, and 45.83% for accuracy, sensitivity and specificity, respectively.

Betala E. et al. [12] projected an SVM and k-Nearest Neighbor (k-NN) algorithm by recording the voice of one afflicted with Parkinson's disease. The features like age and gender are selected for the experiments. For evaluation purpose, the voice recording based on age and gender is taken at three months interval. In order to distinguish significant difference in voice during the evaluation time, SVM method is utilized to calculate UPDRS value.

A. Tsanas et al. [13] anticipated the speech signal algorithm for nonlinear signal method with a huge dataset. The algorithm was implemented in addition with classification algorithm, and support visibility of repeated, isolated, less cost, precise UPDRS tele-monitoring with respect to speech examination.

A. Sharma et al. [14] planned and executed ANN, pattern recognition and SVM so as to help experts to predict the Parkinson's disease. The results say around 85% accuracy for predicting the disease from the voice signal of one with Parkinson's disease while comparing with the normal person.

Shahbakhi et al. [15] offered genetic algorithm and support vector machine for categorizing normal people from people with Parkinson's disease. Based on the fourteen main features of the voice signals, the accuracy obtained is 94.50, 93.66 and 94.22 for 4, 7 and 9 features correspondingly.

Bocklet et al. [16] suggested the automatic detection of Parkinson disease using verbalization, tone, prosodic evaluations based on SVM and Correlation base classification. Experimental results show 90.5% recognition rate.

Cam M. et al. [17] implemented a parallel distributed two-layered neural network, enhanced by filtering and mainstream voting scheme to differentiate people with normal vocal signals from people with Parkinson's disease. To achieve enhancement, training and testing stage are observed, and is known to have the accuracy of greater than 90%.

Kapoor T. et al. [18] implemented Mel-frequency cepstral coefficients to investigate frames in voice signal to recurrent domain, and Vector Quantization to calculate the lowest deformation to recognize voice data. The evaluated results obtained very good enough to predict the Parkinson's disease.

6.3 PROPOSED METHODOLOGY

a. **Restricted Boltzmann Machine**

Restricted Boltzmann machine is an improved, organized neural network, which contains the network of neurons and where each neuron, while activating, will hold approximately random behavior. The RBM consists of visible and hidden layer of neurons [19,20]. RBM is represented by bipartite graphs, where each neuron in the visible layer is interconnected with all the neurons in hidden layer, and there should not be any connection between the neurons in visible layer. Likewise, there should not be any linking between the neurons in hidden layer. A bidirectional and symmetric connection is established among the neurons with same weightage during the training and testing phase.

A graphical portrayal for RBM is shown in Figure 6.3, in which the neurons v_o, v_1, v_2, v_3 belong to the visible layer and the neurons h_0, h_1, h_2 belong to the hidden layer.

RBM is processed and constructed on an energy function defined by input and hidden neurons. The main goal of the learning process in RBM is to reduce the energy cost, which means that the overall energy function described by the RBM should be minimized.

Geoffrey [21] has stated that in RBM, the energy function of the units available in both visible and hidden layer is distinct as

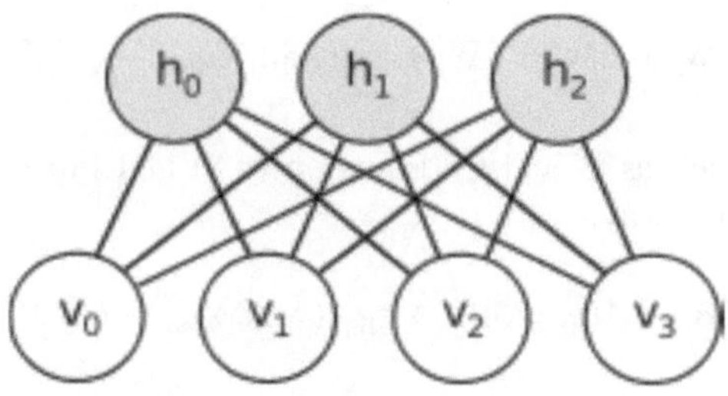

FIGURE 6.3 Graphical depiction of an RBM.

$$E(v, h, \theta) = -\sum_{i=1}^{I}\sum_{j=1}^{J} W_{ij}v_i h_j - \sum_{i=1}^{I} B_i v_i - \sum_{j=1}^{J} C_j h_j \quad (6.1)$$

In Equation (6.1), v represents visible unit, h represents the hidden unit, $\theta = (W, B, C)$ represents the bias value where W interacts symmetrically between v and h, W is the weightage that ranges between visible units i inside v in addition to hidden units j inside h, B and C is bias value of v and h, respectively, I, J are number of units in visible and hidden layers.

Joint distribution of RBM given by energy function is shown in Equation (6.2)

$$p(v, h, \theta) = \frac{1}{Z}e^{-E(v,h,\theta)} \quad (6.2)$$

In Equation (6.2), the Z is partition function or normalization factor is shown in Equation (6.3)

$$Z = \sum_{v,h} e^{-E(v,h,\theta)} \quad (6.3)$$

Visible layer probability, v, is represented in Equation (6.4)

$$p(v) = \frac{1}{Z}\sum_{h} e^{-E(v,h,\theta)} \quad (6.4)$$

The derivative of log probability for the training data based on weightage is shown in Equation (6.5)

$$\frac{\partial \log p(v)}{\partial W_{ij}} = \langle v_i h_j \theta_j \rangle_{data} - \langle v_i h_j \theta_j \rangle_{recon} \quad (6.5)$$

The angle brackets are used to convey the opportunity beneath the distribution specified by the consequent subscript, which leads to straightforward learning rule to carry out stochastic steepest ascent based on log probability of the training data:

$$\Delta W_{ij} = \varepsilon_{W_{ij}}(\langle v_i h_j \theta_j \rangle_{data} - \langle v_i h_j \theta_j \rangle_{recon}) \quad (6.6)$$

$$W_{ij} = \delta W_{ij} + \Delta W_{ij} = \delta W_{ij} + \varepsilon_{W_{ij}}(\langle v_i h_j \theta_j \rangle_{data} - \langle v_i h_j \theta_j \rangle_{recon}) \quad (6.7)$$

Other parameters such as B and C (Equation (6.8) and Equation (6.9)) are also calculated in the same way.

$$B_{ij} = \delta B_{ij} + \Delta B_{ij} = \delta B_{ij} + \varepsilon_{B_{ij}}(\langle v_i h_j \theta_j \rangle_{data} - \langle v_i h_j \theta_j \rangle_{recon}) \quad (6.8)$$

$$C_{ij} = \delta C_{ij} + \Delta C_{ij} = \delta C_{ij} + \varepsilon_{C_{ij}}(\langle v_i h_j \theta_j \rangle_{data} - \langle v_i h_j \theta_j \rangle_{recon}) \quad (6.9)$$

Where, δ is the momentum which is used to increase the learning process when the contrastive divergence objective function is much steeper and narrow and learning rate is represented by ε.

To obtain the impartial sample data of $\langle v_i h_j \rangle_{data}$ is not a tough job, given no direct link among hidden units in RBM. For any arbitrarily chosen training data, v, the binary state, h_j, of each hidden unit, j, is fixed to probability 1.

$$p(h_j = 1|v) = \sigma(B_j + \sum_i v_i W_{ij}) \tag{6.10}$$

$$p(v_j = 1|h) = \sigma(W_i + \sum_j h_j W_{ij}) \tag{6.11}$$

(Geoffrey 2002) presented a quick learning process and subsequently all the binary states of the hidden units are calculated concurrently by Equation (6.10). The amendment in a weight is provided by Equation (6.12):

$$\Delta W_{ij} = \in(\langle v_i h_j \rangle_{data} - \langle v_i h_j \rangle_{recon}) \tag{6.12}$$

The learning rule should equal the exchange objective function by the new name: contrastive divergence (Geoffrey 2002), and it is characterized by the discrepancy between two Kullback-Liebler divergences; by a small disturbance in the objective function.

6.4 IMPLEMENTATION

a. **The Proposed Methodology**

Here, the proposed methodology is described with appropriate algorithms and methods. The diagrammatic representation of the proposed methodology is given in Figure 6.4.

The gathered input data is normalized based on min-max normalization. The normalized input data are fed to the RBM to predict the Parkinson's disease and provides the result in two classes: namely, severe and non-severe.

6.4.1 Data Collection

In this proposed model, Parkinson's Telemonitoring Voice Data Set is utilized based on the UCI Machine Learning Repository. More than 40 patients were sampled with parameters like patient number, patient age, patient gender, time intermission and voice measures. There are more than 5,000 voice footages of the patients and dataset is of CSV format. It has 200 recorded data from each patient.

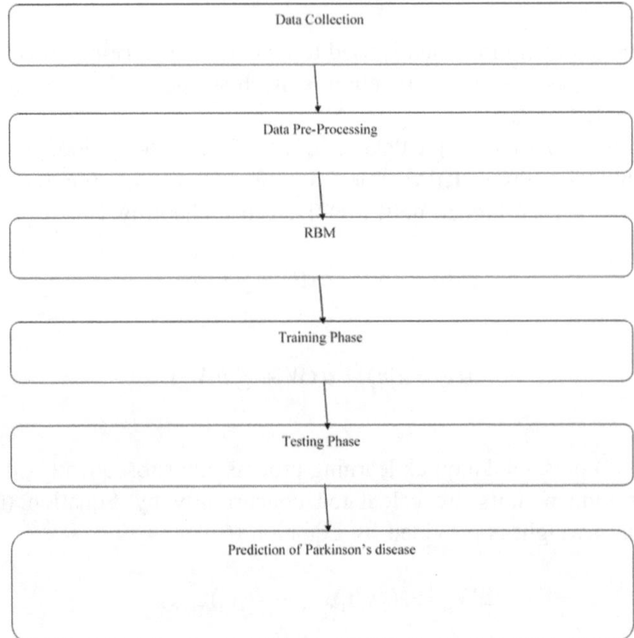

FIGURE 6.4 Block diagram of the proposed approach.

6.4.2 Data Preprocessing

The dataset is normalized from 0 to 1 with the help of min-max normalization approach.

$$\text{Normalized Value of } x = \frac{x - \min(x)}{\max(x) - \min(x)} \tag{6.13}$$

Here, x is the column value, min(x) represents the minimum value for that column and max(x) is the maximum value.

To provide randomness, input values are shuffled and fed into the training model based on lambda function. After, based on the input values, the data and weights of each layer are processed in the training model so as to evaluate the function efficiently and predict the Parkinson's disease.

6.4.3 Deep Learning Algorithm with RBM

After completion of data collection and data, the input values are fed to the restricted Boltzmann machine.

The RBM assembled for the proposed methodology is depicted in Figure 6.5. In order to perform the summarization process, RBM is employed in two phases namely training phase and testing phase. In the training phase, RBM should be

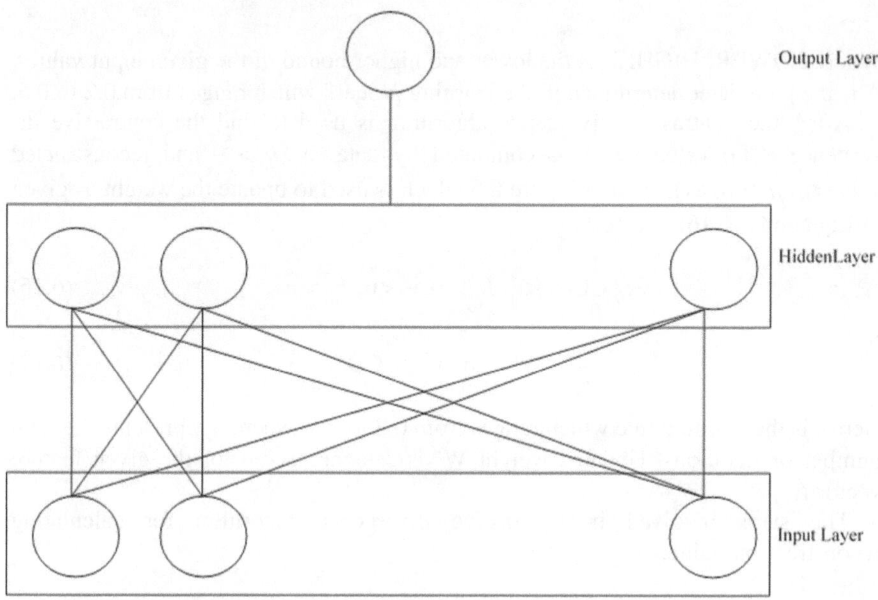

FIGURE 6.5 Block diagram of restricted Boltzmann machine.

trained with the known values. In Second phase, the RBM should be tested for the unknown values.

6.4.4 Training Phase

The restricted Boltzmann machine is trained to learn from the training dataset in the training phase. Based on the energy function, the RBM score is generated for all the sentences. According to Yoshua [22], to train the RBM contrastive divergence learning algorithm is used.

Here, v_i and h_j represent the visible layer neurons and hidden layer neurons. The feature vector fv_{ij} is availed as input towards the visible layer neurons v_i. Hidden layer neurons h_j is calculated based on (6.14).

$$h_j = F(\sum(v_i * W_{ij} + \mathrm{N}(0, \mathrm{Sigma}))) \quad (6.14)$$

Where v_i represents the neurons in the input layer or visible layer, W_{ij} represents the weights between the visible and hidden neurons in the matrix form of i x j, N(0, sigma) represents the normal distribution to have zero mean and unit standard deviation; the non linear function F is represented by Equation (6.15).

$$\mathrm{F} = \mathrm{LOWER} + (\mathrm{HIGHER} - \mathrm{LOWER})/(1.0 + \exp(-\mathrm{A} * h_j)) \quad (6.15)$$

Where LOWER, HIGHER is the lower and higher bound of the given input values, A is the parameter determined in the learning process which ranges from 0.2 to 0.5.

After, the contrastive divergence algorithm is used to find the contrastive divergence (CD) value which is computed by data $<v_i, h_j.>$ 0 and reconstructed value$<v_i, h_j.>$ n, as shown in Figure 6.6 which is used to update the weight as given in Equations (6.16) and (6.17)

$$CD = <v_i, h_j> 0 - <v_i, h_j> n \tag{6.16}$$

$$W' = W + \varepsilon * CD \tag{6.17}$$

Here ε is the learning rate which ranges from 0.2 to 0.5, where n represents the total number of iterations. Updated weight W' is sentence score for the given feature vector $fv_{i,j.}$

The steps involved in contrastive divergence algorithm for calculating reconstruction value:

Step 1. Assign input data to the visible neurons v_i

Step 2. Calculate h_j for each hidden neuron and obtain the $<v_i,h_j.>$ 0 value.

Step 4. Calculate h_j using v_i from Step 3 and compute $<v_i,h_j.>$ 1 value as shown in Figure 6.4.

Step 3. Now compute v_i value based on the obtained h_j value in Step 2 which is called reconstruction

Step 5. Repeat n times steps 3 and 4 to calculate $<v_i,h_j.>$ n.

The overall algorithm for RBM is given below:

Step 1. For each input value do:

Step 2. Assign matrices value as $CD_i = 0$, $CD_j = 0$, i.e., weights between visible and hidden layer

Step 3. Carry out steps 1 to 5 in the algorithm for calculating the reconstruction value.

Step 5. Accumulate $CD_j = CD_j + <v_i,h_j.>$ n based on the updated values of hidden layer h_j

Step 6. Calculate average of CD_i and CD_j

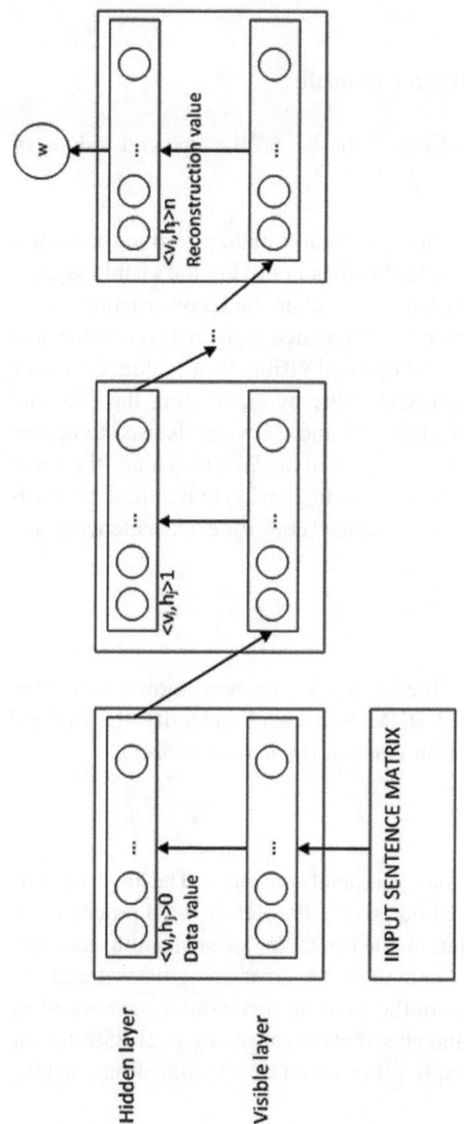

FIGURE 6.6 Block diagram of computing reconstruction value.

Step 7. Calculate CD=$CD_i - CD_j$ (i.e, <v_i,h_j.> 0-<v_i,h_j.> n)

Step 8. Update weight W' = W + ε ∗ CD

Step 9. Compute error function using sum of squared difference between v_i in 1 and v_i in n

Step 10. Repeat the above steps until error is small.

Step 4. Accumulate $CD_i = CD_i$ + <v_i,h_j.> 0 based on the updated values of visible layer v_i

First, assign the input value to the visible layer v_i, then initialize zero to the matrix values CD_i and CD_j which represent the weight values involving the visible layer v_i and hidden layer h_j. Based on given input value, calculate the reconstruction value. Accumulate the <v_i,h_j.> 0 in CD_i based on the updated hidden layer value and accumulate the <v_i,h_j.> n in CD_j based on the updated visible layer value. Calculate the average for CD_i and CD_j and find the CD value by subtracting the CD_i and CD_j(i.e,<v_i,h_j.> 0-<v_i,h_j.> n). Now, the product of ε and CD value is used to update the weight W and the updated weight is represented in W'. Calculate the error function using sum of squared difference between v_i in 1 and v_i in n until the error is small. Finally, the updated weight W' is the sentence score for each sentence based on the given feature vector values.

6.4.5 Testing Phase

Here, the RBM is assigned with unidentified input values and known parameter values to hidden units. Now, the trained RBM will clearly identify the optimal features and provide the accurate prediction class in the testing phase.

6.5 RESULT ANALYSIS

Result analysis of the proposed methodology is analyzed here. The investigation about the projected methodology is carried out giving the step-by-step procedure of the methodology. The computed evaluation measures for classification accuracy based on total UPDRS score is 98.826% and 84.562% for training dataset and test dataset, respectively. While comparing with the existing methodology proposed by Nilashi et al. [23] and Srishti et el. [24], the classification accuracy is 21.85% for the total UPDRS score. The comparison graph value based on classification for total UPDRS score is shown in Figure 6.7.

The computed evaluation measures for classification accuracy based on motor UPDRS score is 90.273% and 86.344% for training dataset and test dataset, respectively. While comparing with the existing methodology, the classification accuracy is shown to be 21.85%. The comparison graph value based on classification is depicted as in Figure 6.8.

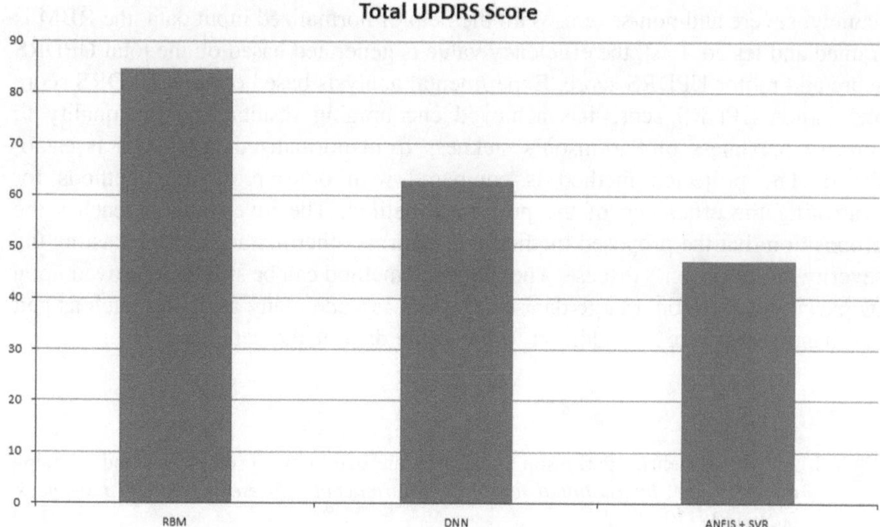

FIGURE 6.7 Comparison graph value based on classification for total UPDRS score.

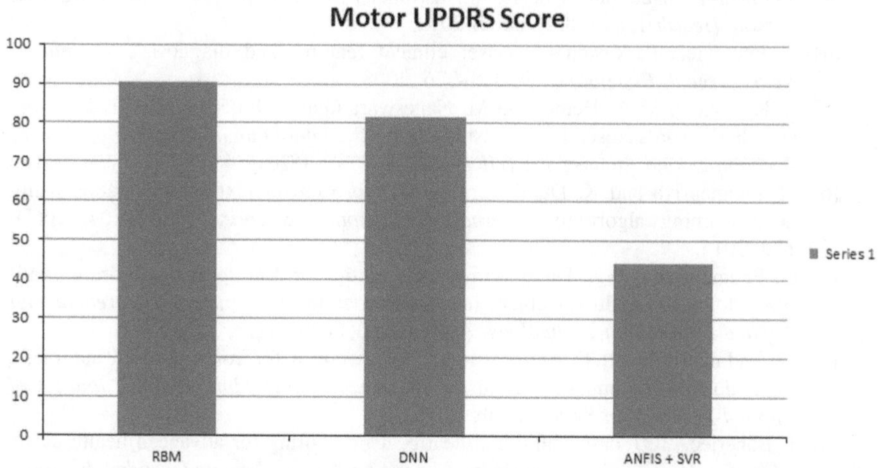

FIGURE 6.8 Comparison graph value based on classification for motor UPDRS score.

6.6 CONCLUSION

The proposed methodology has implemented a deep neural network using RBM to forecast the severity of Parkinson's disease. To perform an efficient forecast of Parkinson's disease, first, the contrastive divergence learning algorithm is utilized to train the RBM which is the basic elementary unit. The gathered input data is normalized based on min-max normalization. The normalized input data are fed to the RBM to predict the Parkinson's disease and it provides the result in two classes:

namely, severe and non-severe. With the help of normalized input data, the RBM is trained and tested. Last, the efficiency value is generated based on the total UPDRS score and motor UPDRS score. Experimental analysis based on total UPDRS score and motor UPDRS score has achieved encouraging results, and the quality to forecast harshness of Parkinson's sickness from normalized input data is established. The projected method is compared with other prevailing methods for evaluating the efficiency of the projected method. The investigation reaches the proposition that the proposed method outperforms other methods in forecasting the severity of Parkinson's disease. The proposed method can be still be improved upon by executing them on a large dataset, that can include many attributes such as gait and handwriting, etc., in addition to the voice data of the patients.

REFERENCES

[1]. A. M. Alshehri, 'Parkinson's disease: an overview of diagnosis and ongoing management', *International Journal of Pharmaceutical Research & Allied Sciences*, 6(2): 163–170, 2017.

[2]. D. H. Ackley, G. E. Hinton, and T. J. Sejnowski, 'A learning algorithm for Boltzmann machines', *Cognitive Science*, 9: 147–169, 1985.

[3]. S.-E. Soh, J. L. McGinley, J. J. Watts, R. Iansek, and M. E. Morris, 'Rural living and health-related quality of life in Australians with Parkinson's disease, *Rural and Remote Health* 12: 2158. (Online) 2012.

[4]. J. Jankovic, 'Parkinson's disease: clinical features and diagnosis', *Journal of Neurosurgeon Psychiatry*, 79: 368–376, 2008.

[5]. S. Kanagaraj, M. S. Hema, and M. Nageswara Gupta, 'Environmental risk factors and Parkinson's disease – A study report', *International Journal of Recent Technology and Engineering (IJRTE) ISSN: 2277-3878*, 7(4S2): December, 2018.

[6]. G. Padmapriya and K. Duraiswamy, 'An approach for text summarization using deep learning algorithm', *Journal of Computer Science*, October 07, 10(1): 1–9, 2013.

[7]. G. Padmapriya and K. Duraiswamy, 'Association of deep learning algorithm with fuzzy logic for multi document text summarization', *Journal of Theoretical and Applied Information Technology*, April 10, 62(1): 166–173, 2014.

[8]. G. Padmapriya, and K. Duraiswamy, 'An approach for concept-based automatic multi document summarization using machine learning', *International Journal of Applied Information Systems*, July 2012.

[9]. I. Banerjee, 'IOT based fluid and heartbeat monitoring for advanced health care', *Classification techniques for medical image analysis and computer aided diagnosis* 4, Chapter 8 (ISBN: 978-0-12-818004-4), 2019.

[10]. P. Mamoshina, A. Vieira, E. Putin, and A. Zhavoronkov, 'Applications of deep learning in biomedicine', In *Proceedings of the American Chemical Society Molecular Pharmaceutics*, pp. 1445–1454, 2016.

[11]. I. Rustempasic and M. Can, 'Diagnosis of Parkinson's disease using fuzzy c-means clustering and pattern recognition', *South East Europe Journal of Soft Computing*, 2(1): 42–49, 2013.

[12]. B. E. Sakara, and O. Kursunb, 'Telemonitoring of changes of unified Parkinson's disease rating scale using severity of voice symptoms', 2014.

[13]. A. Tsanas, M. A. Little, P. E. McSharry, and L. O. Ramig, 'Nonlinear speech analysis algorithms mapped to a standard metric achieve clinically useful

quantification of average Parkinson's disease symptom severity', *Journal of the Royal Society Interface*, 8(59): 842–855, 2011.

[14]. A. Sharma, and R. N. Giri, 'Automatic recognition of Parkinson's disease via artificial neural network and support vector machine', *International Journal of Innovative Technology and Exploring Engineering (IJITEE)*, 4(3), 2014.

[15]. M. Shahbakhi, D. T. Far, and E. Tahami, 'Speech analysis for diagnosis of Parkinson's disease using genetic algorithm and support vector machine', *Journal of Biomedical Science and Engineering*, 7(14): 147–156, 2014.

[16]. T. Bocklet, E. Noth, G. Stemmer, H. Ruzickova, and J. Rusz, 'Detection of persons with Parkinson's disease by acoustic, vocal, and prosodic analysis' In Automatic Speech Recognition and Understanding (ASRU), *IEEE Workshop*, pp. 478–483, 2011.

[17]. M. Can, 'Diagnosis of Parkinson's disease by boosted neural networks', South East Europe, *Journal of Soft Computing*, 2(1), 2013.

[18]. T. Kapoor, and R. K. Sharma, 'Parkinson's disease diagnosis using mel-frequency cepstral coefficients and vector quantization', *International Journal of Computer Applications*, 14(3): 43–46, 2011.

[19]. S. Bind, A. K. Tiwari, and A. K. Sahani, 'A survey of machine learning based approaches for Parkinson disease prediction', *International Journal of Computer Science and Information Technologies*, 6(2): 1648–1655, 2015.

[20]. T. V. S. Sriram, M. V. Rao, G. V. S Narayana, and D. S. V. G. K. Kaladhar, 'ParkDiag: a tool to predict Parkinson disease using data mining techniques from voice data', *International Journal of Engineering Trends and Technology (IJETT)*, 31(3), January 2016.

[21]. G. E. Hinton, 'Training products of experts by minimizing contrastive divergence', *Neural Computation*, 14(8): 1711–1800, 2002.

[22]. Y. Bengio, P. Lamblin, D. Popovici, and H. Larochelle. Greedy layer-wise training of deep networks. In B. Scholkopf, J. Platt, and T. Hoffman editors, *Advances in Neural Information Processing Systems 19 (NIPS'06)*, pp. 153–160. MIT Press, 2007.

[23]. M. Nilashi, O. Ibrahim, and A. Ahani, 'Accuracy improvement for predicting Parkinson's disease progression', *Scientific Reports*, 6, 34181, 2016.

[24]. S. Grover, S. Bhartia, Akshama, A. Yadav, and K. R. Seeja, 'Predicting severity of Parkinson's disease using deep learning', *Elsevier Proceedings of International Conference on Computational Intelligence and Data Science*, 2018.

7 Non-uniform Data Reduction Technique with Edge Preservation to Improve Diagnostic Visualization of Medical Images

Dr. K. Vidhya[1], *Dr. T. R. Ganesh Babu*[2], *Dr. B. Thilakavathi*[3], *S. Poovizhi*[4], *and Dr. Madhumathy P.*[5]

[1]Professor, Department of Electronics and Communication Engineering, Saveetha School of Engineering, SIMATS

[2]Professor, Department of ECE, Muthayammal Engineering College

[3]Professor, Department of Electronics and Communication Engineering, Rajalakshmi Engineering College

[4]Research Scholar, Anna University

[5]Professor, Department of Electronics and Communication Engineering, Dayanand Sagar Academy of Technology and Management

CONTENTS

DOI: 10.1201/9781003194415-7

7.1 INTRODUCTION

The aim of the data reduction method in medical applications is to compress the size of the image without losing medically sensitive information. As a result, creating an effective compression scheme is a continuing problem in the medical industry.

Coding approaches such as spatial domain technique and transform domain technique can all be used to minimize data [1]. A combination of spatial domain procedure and coding approaches to work on the image pixels is followed. Image transformations are used for pixel decorrelation in the transformation domain technique, and information packed coefficients are quantized and coded. The most commonly used transform techniques are DCT, Discrete Cosine Transform, and DWT, Discrete Wavelet Transform. The most commonly used coding techniques are Huffman coding and Arithmetic coding in entropy [2].

The image is divided into sub-images using the Joint Photographic Experts Group (JPEG) compression standard, based on Discrete Cosine Transform (DCT) technique. The sub-images are subjected to DCT, and the resulting coefficients are quantized and coded [3]. One of the drawbacks of DCT is the creation of blocking artefacts.

Wavelet transform-based methods have been shown to be superior to DCT-based methods in terms of providing high compression rates while preserving reasonable image quality [4,5].

Following the implementation of DWT, a wide range of wavelet-based compression methods on the works are carried out. Embedded Zero Tree Wavelet (EZW), which uses parent-child dependencies among sub-band coefficients at the sub-band level, is one; and Set Partitioning in Hierarchical Trees (SPIHT), which uses self-similarity between sub-bands of a wavelet decomposed image, is another; while JPEG2000 algorithm [6–8], which uses Embedded Block Coding with Optimized Truncation (EBCOT), is a third commonly used approach.

Initially the researchers mentioned that compression ratio of 9:1 can be restored for diagnosis of medical images. The effect will not degrade the visual quality. Sone et al. [9] suggest that the compression ratio is accepted for 10:1.

A combination of without-loss and with-loss techniques based on neural network vector quantization and Huffman coding with a compression ratio between 5 and 10 is evaluated. Chen [10] suggested an algorithm based on the SPIHT algorithm for medical image compression. To conduct sub-band decomposition, an 8 x 8 DCT method is used. The data is then structured using SPIHT. This is a block-based strategy with a higher degree of difficulty. The author demonstrated that certain modifications are needed for its application in the medical domain in order to eliminate blocking objects [10].

To search for zero crossing in the second derivative images, and to locate zero edges, Laplacian-based edge detection method is used. To identify the second derivative and locate the edges, Laplacian detector is chosen. On applying Laplacian of Gaussian, the noise is removed by smoothing of edges. But, Laplacian detectors are unable to find the edge orientation. To overcome this issue, Canny Detector [11,12] is proposed, which is used to detect the edges with high speed and better accuracy [13]. Canny edge detector shows decrease in performance for noise-laden images.

The wavelet transform analyzes a set of scaling coefficients and wavelet coefficients, respectively. In two dimensional matrices, wavelet transform is applied row-wise and, then, column-wise. It creates four different sub-bands namely LL, HL, LH and HH. The edge information has to be obtained from the sub-band. To preserve edges to achieve a resultant image with a higher content of visual information, edge detection is frequently used to extract the edge information [14]. This paper shows how to yield image with optimum CR without sacrificing diagnostic information and also with high data level fidelity.

7.2 METHODOLOGY

The medical images are decomposed into low-frequency and high-frequency components using the Daubechies wavelet filter. Because of its high energy compaction, this wavelet is superior in providing high-quality reconstructed images. Using one-dimensional wavelet decomposition in the image, the proposed data reduction algorithm is developed.

To get the appropriate coefficient for data reduction, the coefficient must be selected by means of thresholding. Data reduction algorithm ought not to have a visual quality for clinical images. To remove this bias in images, the output metric is used as the quality parameter for data reduction. Original image x(i, j) of size M × N pixels is associated with reconstructed image $x_R(i, j)$ of size M × N pixels to calculate PSNR, as shown in Equation (7.1).

$$\text{PSNR in dB} = 10 \log_{10}(255^2/\text{MSE}) \tag{7.1}$$

The study is proposed in order to create images with the highest visual quality.

To start with we decompose the original image using Daubechies wavelet. DWT is determined for a collection of digital filters. On decomposing the image, we obtain different frequency sub-bands using wavelet transform.

To extract high level information, a low pass filter is used, and for detailed information, a high pass filter. Filtering the image gives a low pass and a high pass image for the row processing, which sub-images are processed by mean-time filter by a factor of two at column processing stage to produce four sub-bands: LL, LH, HL and HH. In the proposed method, in one-stage decomposition of DWT, the sub-band coefficients and detailed coefficients are obtained. Since errors are more likely to occur at the edges than in the uniform areas, the original edges of the image are retained. In the reconstruction process, the edge information of the wavelet transform sub-band is given more weightage.

The wavelet transform is essentially a complication procedure in signal processing. The difficulty in passing an image through a low pass filter is similar to passing the image in high pass filter. The vertical edges of the original image are represented by the LH sub-band, the horizontal edges by the HL sub-band, and the diagonal edges by the HH sub-band.

An edge image is generated using LH, HH and HL coefficients as follows: The corresponding element $e_{m,n}$ in the edge image [14] is given by Equation (7.2)

$$Em, n = \sqrt{V_{mn}^2 + d_{m,n}^2 + h_{m,n}^2} \tag{7.2}$$

where $V_{m,n}$ is LH sub-band element, $d_{m,n}$ is HH sub-band element and $h_{m,n}$ is HL sub-band element.

$E_{m,n}$ value is calculated for all the sub-band coefficients which leads to the formation of sub-band improved edge coefficients along the other side of less-dominant smooth area coefficients.

The decoded edge information replaces the decompressed image edges. Under LL sub-band, edge image is stored and the sub-band coefficient of LH, HL and HH are fixed to zero when detecting edges. The inverse wavelet transform is used to get the desired edge image using sub-band. The original image edge pixels are removed. Edge information is used as support information to maintain proper image vision quality.

The flattened edges of the boundary are shown in the Beizer curve. The morphological method to detect the boundary fails in some cases. For the same case, wavelet transform technique detects the boundary accurately, thus enhancing the edges of the image.

The edge knowledge is made up of a sequence of zeros with identical values, as well as edge pixels. For processing the edge information, the proposed method employs run length coding. This is the simplest compression technique available for data compression, mainly for long and repeated characters. The edges of the images are replaced with decoded edge information in data reduction algorithm for obtaining good quality image. As a result, image edges are seen in the output image.

The proposed procedure [15] is mainly focused on the process of original image. Original image is decomposed using Daubechies wavelet algorithm.

Based on the incidence of each coefficient, the threshold is calculated in the approximation sub-band. On decomposing the images, threshold approximation coefficient is commonly used. The importance of the approximation coefficients is determined by comparing them to the determined threshold.

Instead of considering only the coefficients equal to the threshold as stated in the previous algorithm, any of the sub-band approximation coefficients with a scale equal to or greater than the threshold is defined as important coefficients. This happens so that the dominant coefficient is selected even if its frequency of occurrence is less and this avoids creation of false boundaries in the reconstructed image.

So, in the proposed method, the sub-band approximation coefficients are larger or equivalent to the threshold coefficients that were selected as important coefficients, and the balance approximation coefficients are retained. Pixel values are reconstructed using IDWT using the significant approximation coefficients chosen by the above procedure. The PSNR is then determined by comparing with original image's pixel information that has been reconstructed.

If required PSNR is obtained. Then, with the significant approximation sub-band coefficients, the image with better visual quality can be obtained. So those coefficients can be further processed. If required PSNR is not obtained, the procedure of identification of significant coefficients are repeated on remaining approximation coefficients. That is, the threshold used to process the remaining coefficients are successively determined based on the next-most frequently occurring coefficient in the approximation sub-band, and the remaining coefficients are tested for significance by comparing with the determined threshold. More coefficients are progressively padded till the condition is met. The significant approximation coefficients that have been chosen are then arithmetic coded.

The compressed image is arithmetic decoded in the reconstruction process, followed by IDWT.

7.2.1 Algorithm of the Data Reduction Algorithm

1. Obtain the input image.
2. Construct matrices with sizes equal to image dimension [SC and SC1].
3. Using Daubechies wavelet, low-frequency and high-frequency components are obtained by decomposing the image into approximation sub-band and information sub-band, respectively.
4. Approximation sub-band important coefficients recognition
   - 4.1. Calculate threshold (Th).
   - 4.2. Compare C and threshold, Th.
   - 4.3. If coefficient C > = threshold, Th, the coefficient is considered important.
   - 4.4. Place them in SC1 where they belong.
   - 4.5. Write down the coefficients that are less than Th.
5. Fill in the SC matrix with coefficients taken from entire sub-bands in the required positions.
6. Find IDWT for framing the spatial pixel values.
7. On comparing with original image and restored pixel values, calculate PSNR.
8. If the PSNR value is about 36 dB, go to step 9.
   Else,
   Repeat the process with the next threshold (Th) to classify relevant coefficients from the remaining approximation coefficients.
9. Build compressed bitstream by arithmetic encoding the significant approximation wavelet coefficients.
10. Decompress the bitstream and decode it.
11. Use IDWT to reassemble the original image.

7.2.2 Algorithm of the Proposed Data Reduction Method with Enhanced Edge Information

1. Read the input image.
2. Extract the original edge pixels.
3. The edge information is encoded.
4. Apply a data reduction algorithm to the original image.
5. Decode the information from the edges.

6. Using the data reduction method, reconstructed edge images are replaced with decode edge information.

7.3 RESULTS AND DISCUSSION

Figure 7.1 illustrates the compressed MRI images obtained on the proposed data reduction algorithm with edge preservation.

MRI image obtained using data reduction algorithm and existing algorithm are shown in Figure 7.1(a)–7.1(d) respectively.

Single level wavelet decomposition is used in the proposed method. The data reduction algorithm with edge preservation and existing algorithm are tested for the 25 MRI images. The average CR of 25 images obtained by using the data reduction algorithm with edge preservation is 5.4 and MICT algorithm has the value 5.8. The PSNR average obtained using data reduction algorithm with edge preservation and existing algorithm are 47.25 dB and 46.12 dB, respectively.

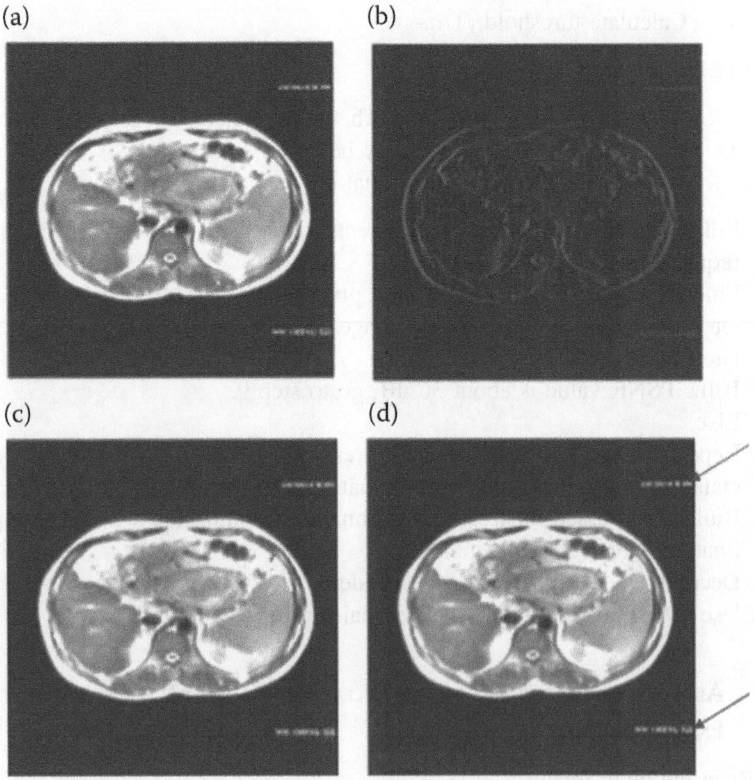

FIGURE 7.1 (a) Original MRI image. (b) Edge information. (c) Reconstructed image using MICT [11] (CR = 6.2, PSNR = 42.14 dB). (d) Reconstructed image using proposed method (CR = 6.0, PSNR = 42.90 dB).

Due to edge data, insignificant rise in compressed bitstream of existing algorithm, or in other words, the value of CR is somewhat reduced in proposed method compared to existing algorithm. Also, the quality of the images is improved on comparing with MICT picture.

On comparing with image edges obtained by MICT and the proposed method, it is obvious that the proposed procedure is greater. The alphabets in MICT algorithm are slightly blurred, whereas the proposed algorithm holds them as clearly as in the original image. Image quality index and MSSIM that supports PSNR are also computed.

The proposed data reduction algorithm with edge preservation retains it as in the original image. On comparison of higher values of PSNR with MICT, preservation of edge information is validated in proposed algorithm.

Figure 7.2 shows compressed CT image with the planned algorithm. The CT image original, edge information and reconstructed images based on proposed data reduction algorithm with edge preservation and existing MICT algorithm are shown in Figure 7.2(a)–7.2(d) respectively. The data reduction algorithm with edge preservation and MICT algorithm are tested for 25 CT images. The CR average is achieved with existing MICT for the value 5.2 and that for the proposed method for the value 5.0, respectively. The PSNR average with MICT has 43.58 dB and that for the proposed method a value of 44.28 dB is attained. The proposed method validates the edge information for CT images.

Figure 7.3 presents compressed X-ray image through proposed method. The X-ray image, edge information and reconstructed images by proposed data reduction algorithm with edge preservation and existing MICT algorithm are shown in Figure 7.3(a)–7.3(d), respectively. The data reduction algorithm with edge preservation and existing algorithm are tested for the 25 X-ray images. The average CR achieved using proposed data reduction algorithm with edge preservation and existing that for MICT algorithm are 6.6 and 6.8, respectively. The average PSNR obtained with proposed data reduction algorithm with edge preservation and MICT are 47.95 dB and 47.67 dB, respectively.

Figure 7.4 presents Ultrasound image compressed with the proposed process. The original Ultrasound image, edge information and reconstructed images using proposed data reduction technique with edge preservation and MICT algorithm are shown in Figure 7.4(a)–7.4(d), respectively. The data reduction algorithm with edge preservation and existing algorithm are tested for the 25 Ultrasound images. The CR average using proposed data reduction algorithm with edge preservation is about 4.8 and that for existing MICT algorithm is 4.9. The PSNR average attained through MICT is 38.35 dB and that for proposed technique is 39.52 dB.

The proposed method employs the extracted edge information to precisely reconstruct the edges. In comparison to MICT, the proposed method produces images with clear edges (Figures 7.1–7.4).

According to the findings, edge preservation results in an insignificant increase in compressed size of the image which results in a penalty for total achievable CRs after comparing with MICT threshold algorithm. However, as seen in the figures, edge protection increases the visual reliability of restored images over MICT images. The method is tested with 20 images under each modality. Table 7.1 shows

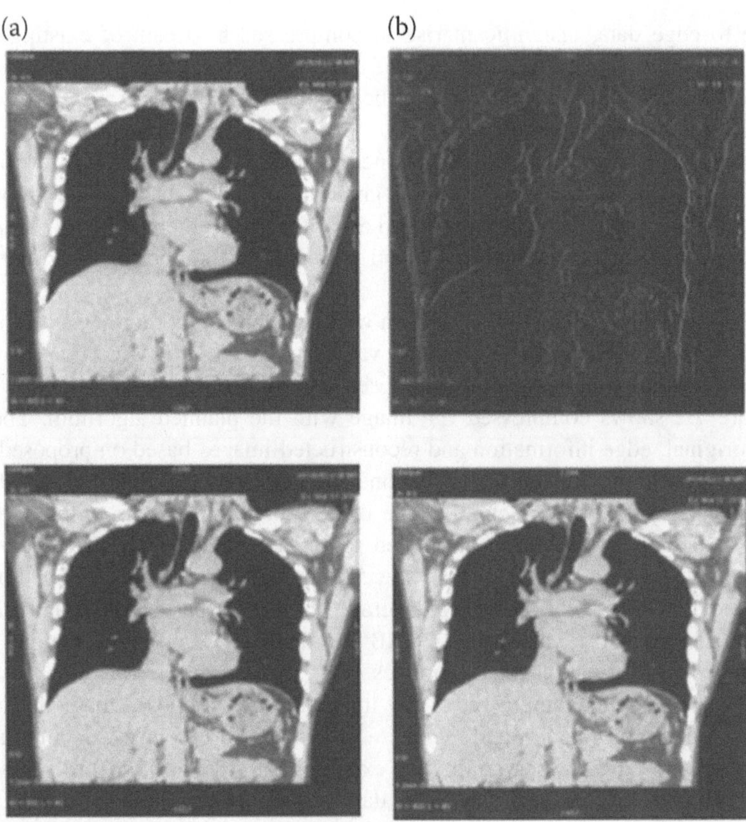

FIGURE 7.2 (a) Original CT image. (b) Edge information. (c) Reconstructed image using MICT (CR = 4.6, PSNR = 42.71 dB). (d) Reconstructed image using proposed method (CR = 4.3, PSNR = 42.97 dB).

CR average and PSNR compared with MICT algorithm vis-a-vis the proposed algorithm for different imaging modalities. Table 7.1 shows that CR is affected by a small margin only on increase in PSNR.

Apart from PSNR, image quality index and MSSIM assessments are also calculated. The average value of image quality index obtained using proposed method for MRI images is 0.84. MICT algorithm average for image quality index is 0.82. This shows that the proposed method is higher on image quality in comparison with MICT.

The average value of Image quality index obtained using proposed method for CT images is 0.87. Using MICT the average is 0.85. Thus, the proposed method has higher value for image quality irrespective of modality.

The average value of Image quality index obtained using proposed method for X-ray images is 0.91. Using MICT algorithm the average is 0.90. Thus, comparatively with MICT and irrespective of picture modality, the proposed method shows higher value.

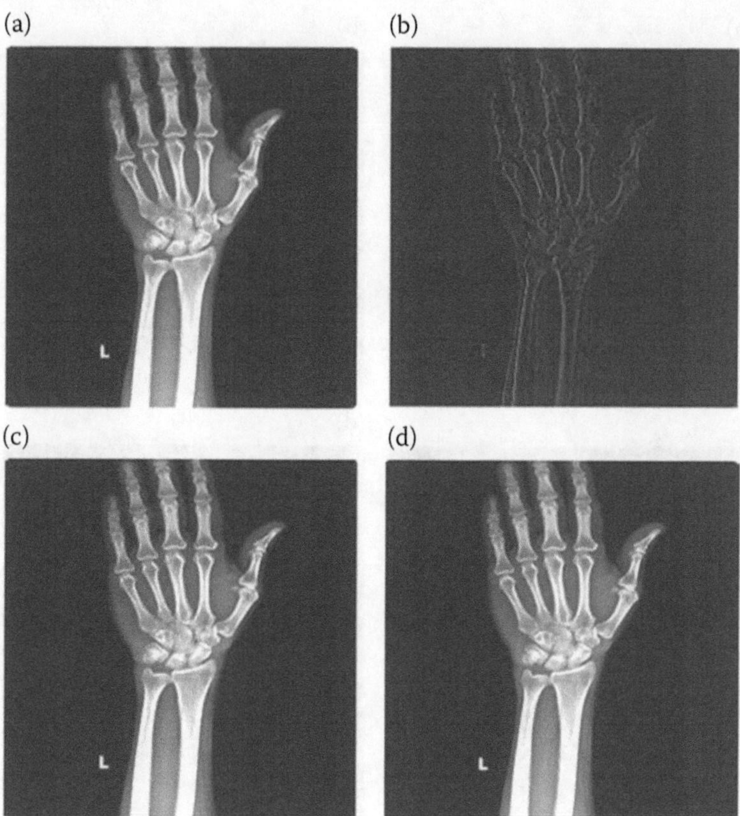

FIGURE 7.3 (a) Original X-ray image. (b) Edge information. (c) Reconstructed image using MICT (CR = 7.3, PSNR = 43.20 dB). (d) Reconstructed image using proposed method (CR = 7.1, PSNR = 43.24 dB).

The average value of Image quality index obtained using proposed method for Ultrasound images is 0.85 which is high compared to MICT with an average of 0.83.

7.3.1 Regression Analysis

The algorithm is further tested with regression analysis. To find the correlation between the two variables, and to predict the value of one variable based on the values of another variable can be achieved by regression analysis. The scatterplot of CR and PSNR for the proposed algorithm technique is generated by using statistical software. The scatterplots are shown in Figures 7.5–7.8 for MRI, CT, Ultrasound and X-ray images. The optimum number of coefficients required for possible maximum compression and also for achieving a PSNR of 36 dB is chosen. So, the minimum value for Y-axis is 36 dB and it will vary above 36 dB depending on the contents of the images considered. Different set of images of the same organ

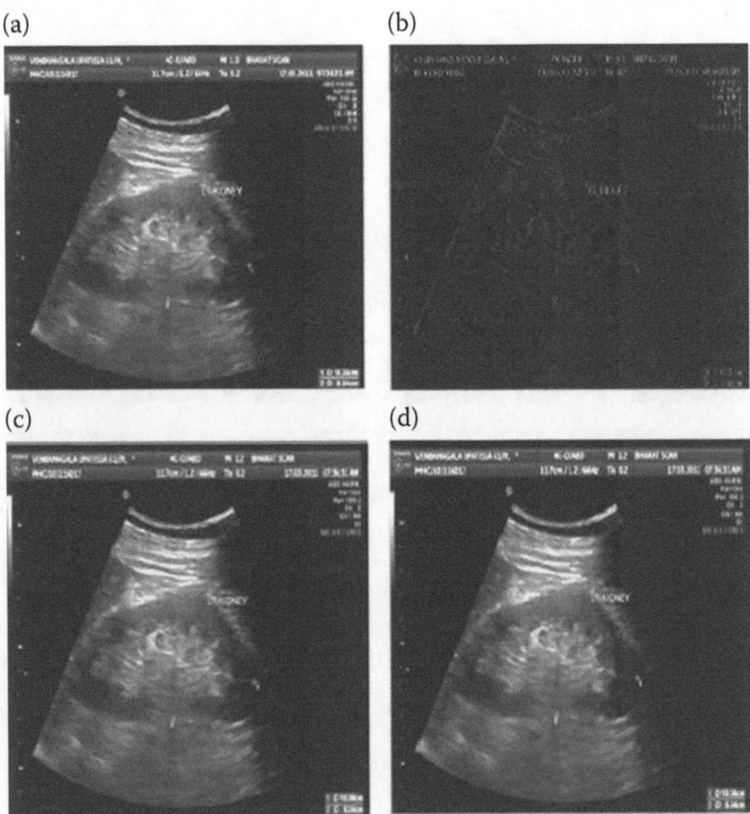

FIGURE 7.4 (a) Original Ultrasound image. (b) Edge information. (c) Reconstructed image using MICT (CR = 5.3, PSNR = 40.70 dB. (d) Reconstructed image using proposed method (CR = 5.1, PSNR = 40.73 dB).

TABLE 7.1

Average CR and PSNR comparison of MICT algorithm with proposed method for different types of images

Imaging Modalities	MICT Algorithm		Proposed Algorithm with Edge Preservation	
	CR	PSNR	CR	PSNR
MRI images	5.8	46.12	5.4	47.25
CT images	5.2	43.58	5.0	44.28
X-ray images	6.8	47.67	6.6	47.95
US images	4.9	38.35	4.8	39.52

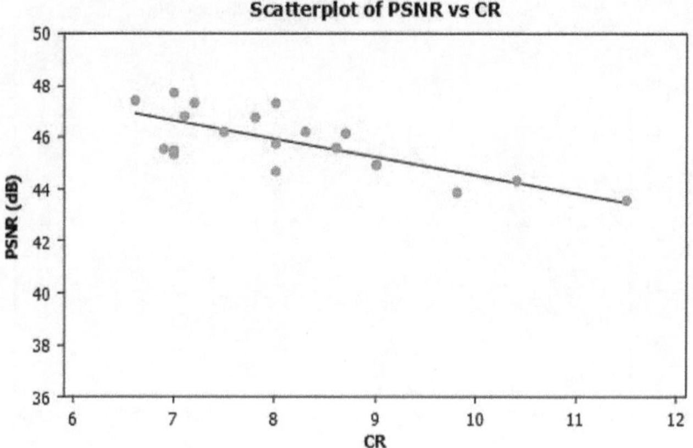

FIGURE 7.5 Scatterplot of PSNR vs. CR for MRI images using proposed method.

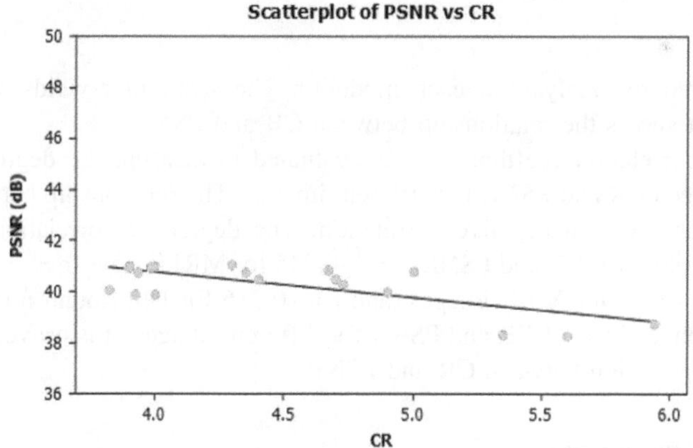

FIGURE 7.6 Scatterplot of PSNR vs. CR for CT images using proposed method.

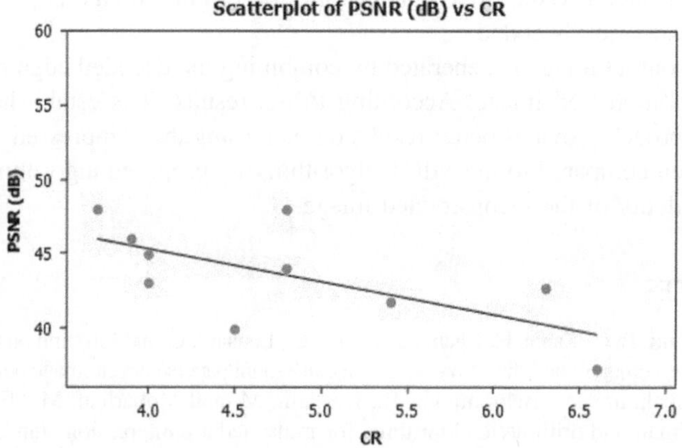

FIGURE 7.7 Scatterplot of PSNR vs. CR for X-ray images using proposed method.

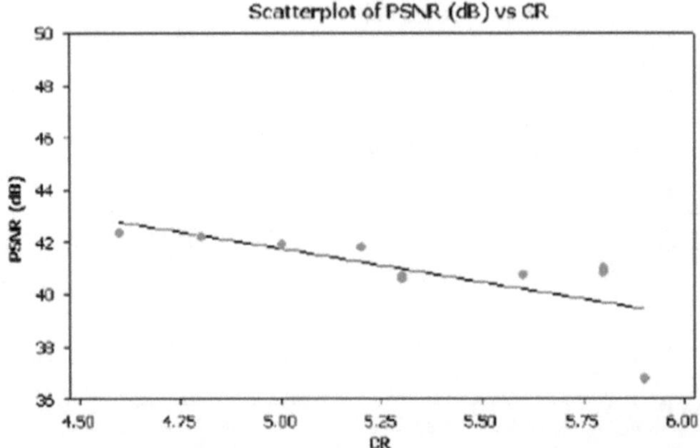

FIGURE 7.8 Scatterplot of PSNR vs. CR for an Ultrasound image using proposed method.

are considered for analysis of each modality. The scatterplot yields a linear regression that shows the relationship between CR and PSNR.

Pearson correlation coefficient (r) is evaluated to measure the degree of association between CR and PSNR for different images. The relationship between these two variables shows a negative correlation. The degree of correlation observed between the obtained CR and PSNR is r = -0.745 for MRI images, r = -0.707 for CT images, r = -0.658 for X-ray images, and r = -0.716 for Ultrasound images. From the regression analysis of CR and PSNR for different images, it is proved that there is modest correlation between CR and PSNR.

7.4 CONCLUSION

Edge detector collects crucial edge information in order to protect the image imperative information. Using MICT algorithm, the original image is processed and the edge information is coded.

The final output image is generated by combining the decoded edge information with the decompressed image. According to the results, it is established that the extended approach provides better results on increasing the compressed sizes of the images. When compared to the MICT algorithm, the proposed algorithm improves the image fidelity of the reconstructed image.

REFERENCES

[1]. Vemuri, B. C., Sahni, F., Chen, F., Kapoor, C., Leonard, C. and Fitzsimmons, J. "Lossless image compression", http://www.cise.ufl.edu/~sahni/papers/encycloimage.pdf, 2007.

[2]. Shahbahrami, A., Bahrampour, R., Rostami, M. and Mobarhan, M. "Evaluation of Huffman and arithmetic algorithms for multimedia compression standards", Int. J. Comput. Sci. Eng. Appl., Vol. 1, No. 4, pp. 34–47, 2011.

[3]. Pennebaker, W. B. and Mitchell, J. L. JPEG still image data compression standard, VNR, New York, 1993.
[4]. Singla, V., Singla, R. and Gupta, S. "Data compression modeling: Huffman and arithmetic", IJCIM, Vol. 16, No. 3, pp. 64–68, 2008.
[5]. Rani, B., Bansal, R. K. and Bansal, S. "Comparison of JPEG and SPIHT image compression algorithms using objective quality measures", in Proc. IEEE International Multimedia Signal Processing and Communication Technologies, pp. 90–93, 2009.
[6]. Taubman, D. S. "High performance scalable image compression with EBCOT", IEEE Trans. Image Process., Vol. 9, No. 7, pp. 1158–1170, 2000.
[7]. Taubman, D. S. and Marcellin, M. W. JPEG2000: Image compression fundamentals, standards and practice, Kluwer Academic Publishers, Boston, USA, 2002.
[8]. Boleik, M., Houchin, J. S. and Wu, G. "JPEG2000 next generation image compression system features and syntax", in Proc. International Conference on Image Processing, Vol. 2, pp. 45–48, 2000.
[9]. Li, F., Sone, S., Takashima, S., Kiyono, K., Yang, Z. G., Hasegawa, M., Kawakami, S., Saito, A., Hanamura, K. and Asakura, K. "Effects of JPEG and wavelet compression of spiral low-dose CT images on detection of small lung cancers", Acta Radiol., Vol. 42, No. 2, pp. 156–160, 2001.
[10]. Chen, Y. Y. "Medical image compression using DCT-based subband decomposition and modified SPIHT data organization", Int. J. Med. Informat., Vol. 76, No. 10, pp. 717–725, 2007.
[11]. Maini, R. and Aggarwal, H. "Study and comparison of various image edge detection techniques", Int. J. Image Process., Vol. 3, No. 1, pp. 1–11, 2009.
[12]. Senthilkumaran, N. and Rajesh, R. "Edge detection techniques for image segmentation – A survey of soft computing approaches", Int. J. Recent Trends Eng., Vol. 1, No. 2, pp. 250–254, 2009.
[13]. Hu, M., Zhang, C., Juan, L. U. and Zhou, B. "A multi-ROIs medical image compression algorithm with edge feature preserving", in IEEE Proc. International Conference on Intelligent System and Knowledge Engineering, pp. 1075–1080, 2008.
[14]. Avijit Sur, A. S., Chakraborty, N. & Saha, P. I. "A new wavelet-based edge detection technique for Iris imagery", Proceedings of the IEEE International Advance Computing Conference, Patiala, India, pp. 120–124, 2009.
[15]. Vidhya, K. "Medical image compression using adaptive subband threshold", J. Electr. Eng. Technol., Vol. 11, No. 2, pp. 499–507, 2016.

8 A Critical Study on Genetically Engineered Bioweapons and Computer-Based Techniques as Counter Measure

Manvinder Sharma[1], *Bikramjit Sharma*[2], *Anuj Kumar Gupta*[3], *Digvijay Pandey*[4], *Joginder Singh*[5], *and Rahul Kakkar*[6]

[1]Assistant Professor, Department of ECE, Chandigarh Group of Colleges

[2]Assistant Professor, Department of ME, Thapar Institute of Engineering and Technology

[3]Professor, Department of CSE, Chandigarh Group of Colleges

[4]Department of Technical Education, IET, Dr. A.P.J. Abdul Kalam Technical University

[5]Professor, Department of Applied Science,Chandigargh Group of Colleges

[6]Associate Professor, Department of Applied Science, Chandigarh Group of Colleges

CONTENTS

DOI: 10.1201/9781003194415-8

8.1 INTRODUCTION AND HISTORY

In today's world, other than natural disasters, we face the real threats of bioweapons and bioterrorism with genetically engineered agents. The process of human intervention to transfer biological organisms to functional genes (DNA) is known as genetic engineering [1]. The manipulation of genes to make new types of genes which have pathogenic characteristics (infectivity, increased survivability, virulence and drug resistance, etc.) is where bioweapons have their use [2]. The "black biology" that is used to create bioweapons is one of gravest threats we face. In the Gulf War of 1991, bioweapons were produced and used for the first time and near about 20 countries engaged in proliferation of such weapons [3]. In the 1970s, genetic engineering began to develop as a heavily researched field and became a global multibillion dollar industry by the 1980s. The knowledge of molecular biology increased exponentially in last decade of the 20th century. Thus, with these genetically engineered pathogens being discovered in R&D laboratories there was the plausibility of theirs being used for many offensive bioweapons. In 1918, the influenza also known as Spanish flu that spreads by coughing and sneezing infected 500 million people (about quarter of world's population at that time) with estimated deaths of 50 million across world. The massive troop movements and living in close quarters during World War I hastened the pandemic [4,5]. In 1979, the outbreak of anthrax, through anthrax spores released in the air was triggered in a military facility which killed 169 people which includes 105 victims. In 2001, anthrax spores were mailed in letters through U.S. Postal service which made 18 people sick with some sort of disease and 5 people died from inhalation of anthrax. In the 2014 outbreak of anthrax, 7 people died and Indian government quarantined 30 houses. In 2016, in another anthrax outbreak, 100 people were infected and 2300 reindeer died in Russia [6].

In 1965, a common cold named B814 came into existence when Tyrrell and Bynoe were studying samples of human embryonic trachea taken from the respiratory tract of an adult. During the same time period, Hamre and Procknow obtained samples from medical students with cold and cultured a virus in tissue culture showing unusual properties that they named Hamre's virus, latter on known as 229E. The relation between B814 and 229E viruses and their familiarity with myxoviruses or paramyxoviruses known at that time was not found to be significant to merit a closer look. In the late 1960s, morphological constraints between certain animal viruses such as mouse hepatitis virus, swine flu, gastroenteritis virus and infectious bronchitis virus, were found to have similarities with some human strains as studied by Tyrrell and a group of virologists working together. Crown like appearance of this new group of virus was officially named as coronavirus [7].

H1N1 is the subtype of influenza A virus which was the cause of influenza in 2009 and has an association with the outbreak of Spanish flu in 1918. During the flu pandemic of 2009, in the United States, a virus was isolated from an infected patient and it was found out that H1N1 virus was made from the genetic elements of four different viruses [8]. From Mexico in 2009 the HIN1 had spread to the United States where it was declared a national emergency by the US President on 25th October 2009. It was already declared a pandemic on 11th June 2009 by World Health Organization (WHO). At least 213 countries were affected by H1N1 pandemic by 21st March 2011 and 16,931 deaths were reported. The preparedness alert of H1N1 pandemic was issued by WHO in 2011 [9].

SARS virus is an animal virus identified in 2003. It was thought to spread from bats and other animals and was perhaps transferred to humans in China in 2002. 26 countries were affected by the SARS epidemic and 8000 cases were reported in 2003. The SARS virus transmission is from person to person. The symptoms of this disease include malaria, headache, shivering, diarrhea, shortness of breath and cough. The countries which were affected by SARS epidemic in 2002–2003 were China, Canada, Singapore, Hong Kong and Vietnam. The vaccines for SARS-CoV is still under development [10,11].

MERS stands for Middle East Respiratory Syndrome. Saudi Arabia first identified this viral respiratory disease in 2012. The typical symptoms of MERS include pneumonia, shortness of breath, cough and fever. 35% of patients have died due to MERS virus from 2012 to till date. 2494 cases were reported by WHO since 2012 with 858 fatalities. Rather than transmitting from human to human, this virus is zoonotic which means the transmission is between human and animals. According to studies, the transmission to the humans is through infected camels [12]. 27 countries have been affected from MERS virus since 2012 including China, France, Bahrain, Egypt, Malaysia, Germany, etc. In Saudi Arabia, 80% of human cases were reported. Most of the outbreaks of MERS virus have occurred in the Middle-East. The vaccines for MERS virus are not currently available and are still under development [13].

The evolution of bioweapons can be categorized in four phases. The first phase was during World War I when gaseous chemicals like phosgene and chlorine was used. The World War II experienced the second phase when use of a cholinesterase inhibitor, nerve agents (e.g. tabun) and plague bombs, and anthrax were used. The third phase constituted 1970 during Vietnam war, a mixture of herbicides stimulated hormonal function was used for defoliation and destruction of crops. Also a new group of "Novichok" and mid spectrum agents like bioregulators, auxins and physiologically active compounds were used. In the fourth phase, genetic engineering and biotechnical evolution are used to generate gene-designed organisms and can be used for a wide variety of bioweapons [4]. Gene-engineered organisms can be used to produce microorganisms with enhanced environmental stability and the aerosol, a microscopic factor producing a venom, toxin or bioregulator, the microorganism which resists antibiotics, therapeutics and routine vaccines, the microorganism with altered immunological profile and the microorganism which can escape detection by antibody sensor system.

The biological weapon is a device with a delivery system which expedites dispersion and appropriate dissemination of the designed biological agent in such a way that it targets the object (human, crop, etc.) with maximum effect. It can be used to inject the virus/bacteria by spraying from airplane on area of denial, or by handheld spray weapon [14].

A minuscule pellet containing ricin is put in a projectile weapon and through the spike of an umbrella, a plant driven toxin is delivered to its aim with targeted deaths in Paris and London in 1978. In biological warfare, small pox virus has been used for long as a lethal weapon. In 1973, to decimate American Indian population, blankets of small pox patients were widely distributed. Also bioweapons can release pathogenic and harmful micro-organisms on lifelines of enemies like killing crops and destroying food reserves of enemies [15]. Anticrop warfare with use of bioweapons results in malnutrition, debilitating famines, food insecurity and decimation of agriculture-based economies. In Vietnam, war defoliants have been widely used as anticrop warfare agents. Wheat smut caused by fungus T. foetida or tilettia caries was used as bioweapon. In 1984, deliberate contamination of salad bars with salmonella typhimrium was done in USA incapaciting voters. In 1995, Aum Shinrikyo released nerve agent (sarin) in Japan sought medical attention of whom 1038 were poisoned and 12 died [16].

8.2 GENETICALLY ENGINEERED PATHOGEN

8.2.1 DESIGNER GENES

A human molecular blueprint was provided by the human genome project which decoded alphabets of life which has complete genome sequence and known for 205 naturally occurring plasmids, 599 viruses, one fungus, 31 bacteria, one plant and two animals. These blueprints enable a manufacturer of bioweapon to make microorganism more harmful and critically life threatening. From the designer genes, the microbiologist can develop synthetic virus, synthetic genes or even a new microorganism. Some of the bacteria develop resistance to antiviral agents or antibiotics; again, identifying antibiotic resistant genes, an organism can be developed which can render resistance to antibodies. For example, genes with beta-lactamase codes defeat action of penicillin [17].

Analogous to natural mutation, entire viruses may be created. By swapping out variant or synthetic genes or by inducing hybridization of viral strains, a new strain of influenza can be created. Altering common virus influenza slightly can make it more deadly. It becomes possible to choose the most lethal characteristics, as the database of microbial genomes with its part list is available. The entire gene could be stitched together with some animal viruses [18] to produce bioweapons.

8.2.2 Binary Bioweapon

To use the form of pathogen, the parts of a two-component system containing innocuous parts are mixed. Multiple plasmids that code for virulence or any special characteristics contained by many pathogenic bacteria are used to create a bioweapon. The virulence of plague, anthrax and other diseases are enhanced by these plasmids and can be transferred often across species barrier to different kinds of bacteria [19]. A virulent plasmid and host bacteria can be produced in required quantity for bioweapons.

8.2.3 Gene Therapy as Bioweapon

Gene therapy is a permanent replacement in genetic composition of any person by replacing faulty gene or repairing it. The two classes of gene therapy involve somatic cell line (therapeutic) and germ cell line (reproductive). Changes in DNA of somatic cell could not be passed onto next generations but only affects individual; however, changes in DNA of germ cell would be followed by next generations. The first genetically altered primate was produced by the virus of jellyfish gene into rhesus monkey egg [20].

8.2.4 Stealth Virus

A cryptic viral infection which enters the human genome and remains dormant for extended time periods is the stealth virus. The triggering of this virus is later done by external stimulants. Through this external stimulus, the virus gets activated and causes a disease. Herpes virus is carried by many humans which can be activated to cause genital lesions or oral lesions. Some people who had chicken pox earlier in life carries vericella virus. Segments of DNA (Oncogenes), when triggered, can initiate misbehavior and wild cellular growth. Some genes can cause cancer [21].

A stealth virus can infect the genome of the population through a bioweapon and can be triggered anytime for a targeted population; or, the government can be blackmailed by the threat of its triggering [22].

8.2.5 Hot Swapping Disease

A zoonotic disease is transmissible to human by a virus in the animal species. These viruses have a natural animal reservoir in which they reside and cause no damage or very little damage to that animal. However, when transferred to human they cause significant disease. Examples of natural animal reservoirs are water fowl for eastern equine encephalitis, birds for west line virus, bat for ebola virus and corona virus, and rodents for hanta virus. HIV virus which is natural reservoir for chimpanzee becomes disease in humans as AIDS [23]. The bioweapon can contain these viruses taken from animals as hot swapping disease.

8.2.6 Designer Disease

Molecular biology has reached a point where symptoms of hypothetical diseases can be proposed, designed and created from these pathogens known as designer disease. These designer diseases in bioweapons can cause a disease by inducing specific cells to multiply or divide rapidly, turning of immune system and programed cell death [24].

8.3 COMPUTER-BASED DETECTION AND COUNTER MEASURE TECHNIQUES

With the development of bio-robots, a computerized artificial system, bio-detection where robotic insects mimic certain biological processes which can help in detection of viruses has been spurred on with new vigor. Single operation tasks like screening of blood samples, DNA processing, identification of genes, the scan of presence of virus/bacteria and monitoring of genetic cell activity can be done through computerized techniques. Bio-sensors, using either electrochemical devices or fiber-optics, are used for detecting microorganisms in food, military application and clinical work. For example, for detection of candida albicans, an immunosensor is used. Again, optical sensor is used to detect bacillus anthracis. In USA, several systems have been developed to detect bioweapons; while to detect biological agents that cause metabolic damage, polyvalent immunosensor has been developed. Combination of neural informational network, electronic nose and laser eyes to detect particle densities with alarm has been developed.

8.3.1 Computer and Artificial Intelligence-Based Counter Measure Techniques

The implementation of the computer-based technology and artificial intelligence in health sector means that the patients who require care are diagnosed and treated using sensors embedded in the smartphones and computers. Since it seems like the diagnosis and treatment are quite simple, the steps to properly treat the patient however need so many unseen background factors to be taken care of like data collection through various modes such as calls and interviews; results to be processed and analyzed; and accurate and proper diagnosis to be done using multiple sources of the data. Next, the treatment method to be chosen is prepared and administered, and last, continuous monitoring of the patient and the aftercare which includes follow-up appointments [25]. Figure 8.1 shows the stages in which the computer technology can be implemented in the healthcare sector and provides a solution to combat various diseases that spread by bioweapons.

There are four solution stages for computer-based technology which are dependent on each other. There is an interconnection between all the four stages of the architecture such that the data from the first stage is processed to the next stage. In the first step, there is data collection from interconnected devices such as sensors by a connected computer. These devices include sensors, monitors, actuators, camera,

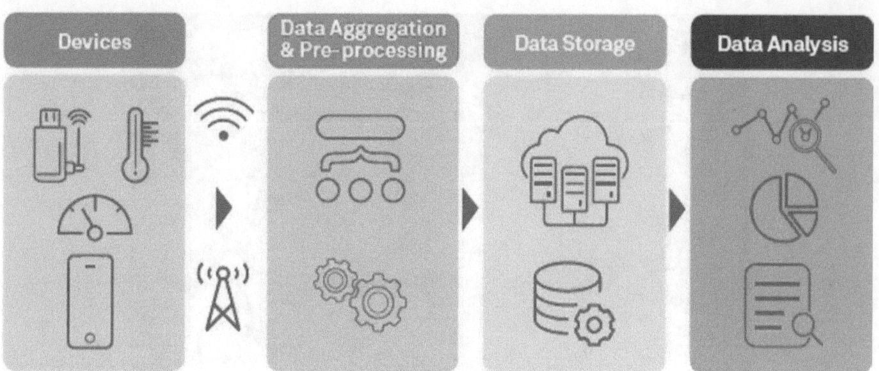

FIGURE 8.1 Solution stages for computer-based technology.

detectors, etc. embedded or connected with computers. In the second step, the pre-processing and standardization of the data takes place which is moved to the cloud network in third step for storage. At the fourth and final stage, the analysis and management of data is done [26].

Scientists of South Korea have trained AI to speedily spot anthrax to combat bio-wars. The holographic microscopy is combined with artificial intelligence to detect anthrax. The last available methods to detect anthrax were after the bacterium infects tissue. Figure 8.2 shows anthrax spores under microscope. Using holographic telescope, the intensity of light scattered off an object can be captured. The researcher took 400 individual spores of 5 different species of bacteria, out of which one is bacillus anthracis (anthrax), and trained the neural network algorithm to detect the spores, and the AI detected the spores [27].

8.3.2 Computer-Assisted Surgery as Counter Measure

Due to the spread of any virus/bacteria during a bio-war, there is need to update the system for detecting agents in bio-weapons by providing computer-assisted surgery (CAS), a surgical process improved by integrating computer technology in the planning and guiding of the surgical intervention. CAS aims to increase the accuracy, reduce invasiveness and costs related to the procedures of surgery [28]. Figure 8.3 shows the pipeline of the computer-assisted surgery.

Acquiring the data of the patient is the first step in the pipeline. This data acquisition is based on imaging such as X-rays, CT scans, MRI and any type of measurement taken from the optical tracking system. The second step is segmentation of data, which is to process the data so that information of greater detail can be extracted. Then, it is the step for data visualization where surgical planning is done. During visualization, the surgeon studies the anatomy of the patient and the surgical procedure is virtually performed by studying the pathology of the patient-specific data acquired in the first step [29]. The surgical outcomes which are predicted are examined in some cases. All these three steps of image acquisition, image processing and visualization comes under pre-operative

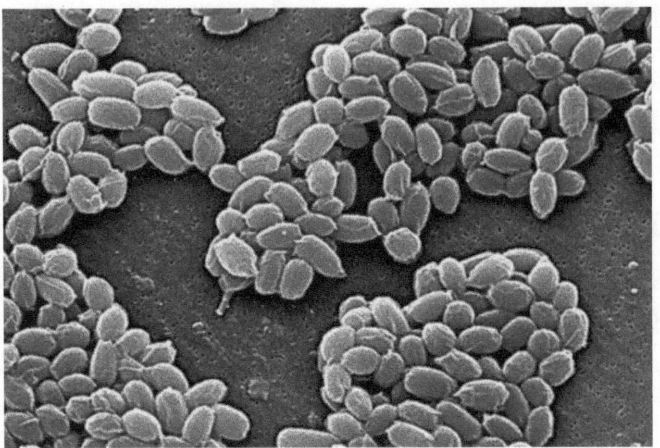

FIGURE 8.2 Anthrax spores at 12,000x microscope.

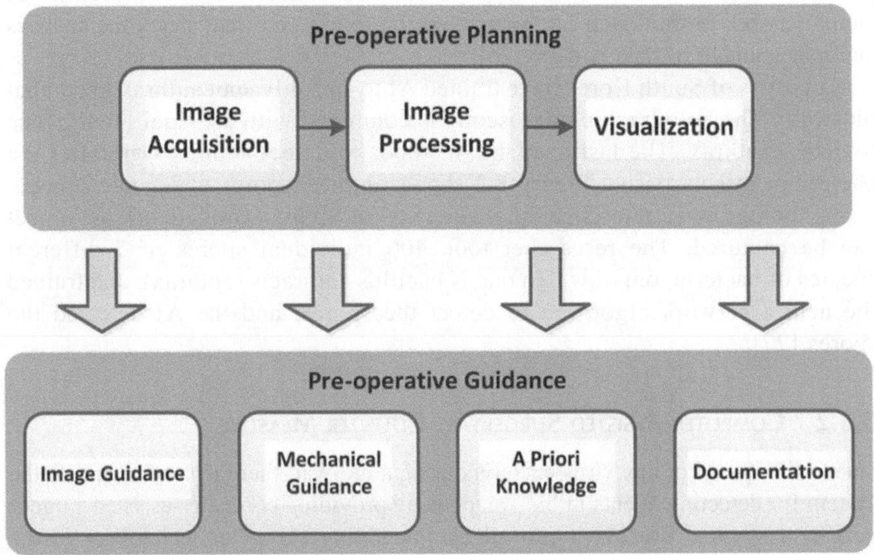

FIGURE 8.3 Pipeline of computer-assisted surgery.

planning. Next, the whole surgical plan is transferred to the operating room so that it can be applied to the process of surgery in actual. The plan is implemented in the surgical process through image guidance, mechanical guidance or with the surgeon's mental model that he had prepared in the planning stage. Image guidance and mechanical guidance are explicit in nature while the documentation is implicit guidance. In computer-assisted surgery, visualization has a very significant role. It comes under the planning phase where the acquired data, predicted outcomes of the surgery and the derived measurements of the patient are displayed [30]. The efficient interaction of the surgeon with various components is also enabled through visualization. Many general tasks are facilitated in computer-assisted surgery for planning and guiding the surgery like spatially understanding the anatomy and pathology, access planning, resection planning, reconstruction planning and implant planning. The patient data visualization is explored by the surgeon which comes under spatially understanding the patient data anatomy in the specific region. This task is very important and basic in complex anatomy [31].

The hand motions of the surgeons are facilitated with the help of computer-assisted surgery. Also known as robotic surgery, it limits the operation space of the surgeons. The various advantages of the devices used in computer-assisted surgery are magnified vision, improved access and stabilization in implementing the instruments. The operative dexterity is reduced significantly in the instruments of standard endoscopy whose degree of freedom is four. The reversed hand motions are required by the operator where the motion of the trocar is dependent on fulcrum. Moreover, the manipulation of operating tips in the shaft shear instrument leads to fatigue in the hand muscle as high forces are induced. Also, in endoscopic surgery there is incompatibility between the skills of human motor visual motor. All these limitations are overcome by enhancing the computer-assisted or robot-assisted surgery. The robotic arms are mounted with microwrist instruments which reproduces the motion in scaled proportions with the help of computer interface. The activity of the wrist is emulated to all the axis of the wrist through the instrument. In confined spaces of operation, the enhancement of dexterity is being transformed from motion scaling and tremor filtering. Figure 8.4 shows a console from where the surgeon operates in 3-D field of operation. It is a robot assisted tele-manipulation system with a camera and two instrument arms [32].

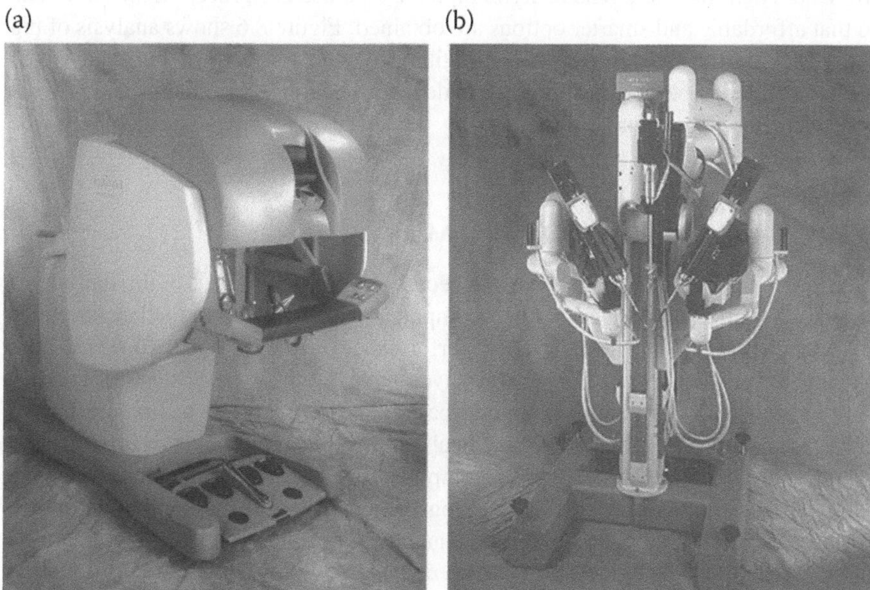

FIGURE 8.4 Computer-assisted tele-manipulation system.

8.3.3 Big Data as Healthcare

The prevention, diagnosis and treatment of human health is the main aim of any healthcare system. Health professionals, facilities and funding companies fall under the healthcare. The health professionals include doctors and nurses, healthcare facilities include clinics and hospitals whereas funding companies are those which run the healthcare facilities. The various health sectors include medicine, psychology, nursing, dentistry to name a few [33]. The patients' medical history in the form of personal history and medical data is required by the doctors at such levels of disease treatment as needed. EHR (Electronic Health Records) are in practice nowadays which store all the necessary information of the patient required for treatment. Figure 8.5 shows big data in health care applications.

The transmission, reception, storage and manipulation of data are all possible with computer-aided technology, given this data is required by the healthcare sectors and departments within the sectors to work efficiently [34].

There are many data care components related to healthcare which might be gathered from the patient; information which results in quality improvement and makes the service efficient by reducing the costs and errors in healthcare and for the medical sectors. Some healthcare-related data components include Medical Practice Management (MPM) and Personal Health Record (PHR). The cost control and health improvement are the factors to be accounted for in the healthcare field which is possible through big data. Various healthcare centers receive the data for further investigation making it confidential. A large amount of information related to diseases and their treatment is possible through the computer-aided technology. Vast amount of data is stored in the data warehouses from different sources through Big Data Analytics [35]. The analytic pipelines are used to process this stored data so that affordable and smarter options are obtained. Figure 8.6 shows analysis of big data. The healthcare sector has been digitalized with the help of big data which further enables important stakeholders to deal with healthcare problems and provide improvement in preventive care, and by the discovery of new medicines in the healthcare field.

8.3.4 Computer-Assisted Decision Making

With every passing day, computers are becoming very significant in the daily lives of a broad section of people. As microcomputers were invented in the late seventies and due to the enhancement in their performance in the eighties, computers are playing a very significant role in our lives even as the whole lifestyle gets revolutionized with computers. Computer-based disease treatment techniques have a great social impact due to their medical field applications. For example, large hospitals are being run depending on computer-based measures and techniques like computerized tomography (CT) and sonography, and where the facilities provided by the computers like data storage and analytics are integral in disease diagnosis and treatment. In medicine field, the computers are majorly used in information system for hospitals, medical imaging, analysis of data in medicine, monitoring of the patients, computer-assisted surgery, decision-making and therapy, treatment

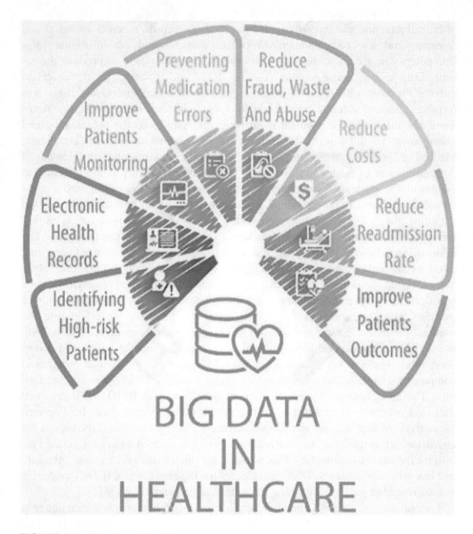

FIGURE 8.5 Big data in healthcare.

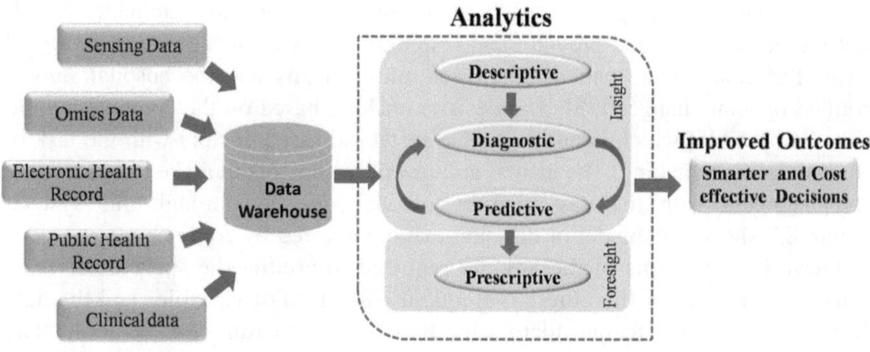

FIGURE 8.6 Analysis using big data.

of critical patients, tele-medicines, electronic health records, medical databases and research, and use of computers in offices and hospital administration [36]. Computers can accept, store and access the data automatically to produce the results. Data in very large amounts can be stored and processed in computers to provide the user with required information as well as computers are itself performing rapid and accurate calculations for users to become more productive. Apart from this, informatics in medical is growing very rapidly as the organization and management of the information for patient care seems to be very significant. Medical informatics seeks to facilitate education and research in biomedical through the use of information networks and computers. Thus, if the information system of the hospital is computerized, then the data can be continuously transmitted, stored and monitored at all transactions, including valuable information regarding the patient care becomes easily accessible. Physicians access all information directly through the application of computer-based techniques [37]. The information system in the hospital, and the various softwares customized according to the needs and requirements of hospitals, covers registration, billing, diet, pharmacy, accounts and biomedical maintenance. Regarding medical research, the data is collected in large numbers after which compilation, analysis and interpretation of the data using various time-consuming statistical methods for calculating the standard deviation, error, t test, Z test and chi square test become necessary. With the aid of computer-based techniques, all these calculations can be done in a very short time. The various statistical packages of good quality include biomedical computer package, statistical package for social sciences (SPSS), Genstat and Epi-Info. The first package in medical sciences developed was BMD and it provided statistical programs at an advanced level. SPSS provides various statistical options for analysis of multivariant and simple statistics. The most powerful package is the Genstst which is used for variance analysis. WHO developed a package called Epi-Info for the study of epidemics. This package can process the word, analyze the data and has graphical abilities. WHO and Center for Disease Control (CDC) made this non-copyrighted package available for statistical programming [38].

Laboratory computing is one of the applications of computer-based techniques in the medical field. The analysis of laboratory tasks includes photometry, blood chemistry, microbiology, to name a few. Proper validation of the results with the patient identification is done. This contributes to the efficiency in the care system of patients. The monitoring machines based on the computers can automatically collect the heart rate, blood pressure and respiratory activity of the patient in digital form. The chart of the patient is updated automatically and the hospital staff is notified of vital changes [38]. The decision-making based on the computers is facilitated by CMD which is an interactive system to assist doctors with the task of clinical decision making. The natural abilities of the doctors to make judgments are complemented with the vast memory of the computer through this system. Figure 8.7 shows the model of decision-making assisted by computers.

Large interventions of therapy are required to predict the survival of critically ill patients. For this, there is frequent collection of variables and the data derived is provided to the doctors for their use in treatment of patients. The recorded data to be significant should be clear in its intended message to

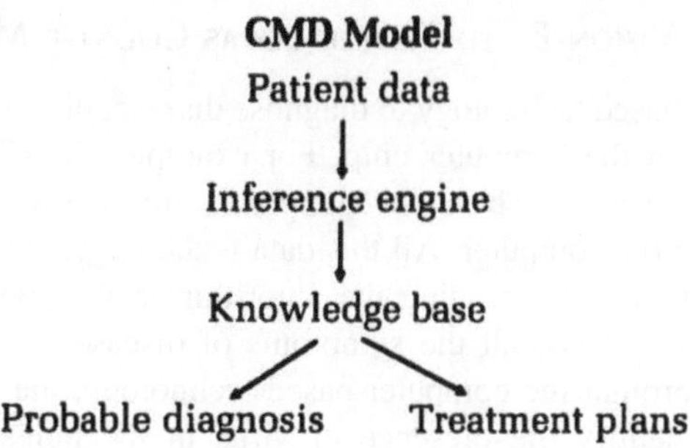

FIGURE 8.7 Decision making model assisted by computers.

differentiate from the large quantity of information collected. The Intensive Care Unit (ICU) computerizes all the necessary and required information so that there is perfect management of the patient's data. For example, there has been the development of closed loop system for vasodilator infusions [39]. Again, the computer technique-based therapy includes various methods to plan, monitor and adjust the dosage of toxic drugs such as in antibiotics, where the regimens of dosage can be planned by the physicians through the target peak selection. Besides, various surgical procedures can be planned, taught and performed with the aid of computer-based technologies, including robotically assisted surgery (RAS) which is one of the major and recent developments in the medical field. The use of computer software and robotic devices allows the surgeons to implement minimal-invasive techniques. The images of the human body created with the technology of medical imaging can be given high resolution with computer-assisted techniques because there is dedicated hardware and software for this purpose so that high resolution images are obtained in CT scans, gamma cameras, ultrasound and Magnetic Resonance Imaging (MRI). This apart, the information system of the hospitals can be integrated with these workstations where various technologies in the field of medicine has helped transform medical sciences completely. These technologies include 3-D printing, Artificial Intelligence, BCI and BBI's, Robotics, Electronic Diagnosis, and technologies related with the interaction between patient and physician. For instance, 3-D printing can be used in future to test the toxicity of a specific drug on the human body. The use of artificial intelligence in the field of medicine includes the evaluation and analysis of patient's personal and biometric data including all the levels of diet and activities. Thus, a whole new understanding level of patient cure can be achieved through Artificial Intelligence. The complex connections of human and computers are incorporated with the advancement of interface between brain and computer. With the advancement in technology, healthcare has become efficient, better, more accurate and easier for patients, physicians and hospital staff [40].

8.3.5 Computer Vision-Based Techniques as Counter Measure

Using the computer-based technology to diagnose disease, the data is stored through the sensors present in the computer chip. For example, the CT scan of lungs is stored in the camera sensor. The other symptoms are also identified through the sensors embedded in the computer. All this data is then aggregated and configured. The configuration is done by running the algorithm in the applications of smartphones and computers. Once all the symptoms of disease are recorded and configured separately through the computer-based technology, the data is then stored and analyzed. To identify the presence of virus in the lungs through CT scan images, many reports of radiology are trained in order to get a better diagnosis. Since the computational techniques have become very advanced, CPUs and GPUs are required which are provided by the cloud in the form of virtual machines. The progressive CT scan images of the lungs are shown in Figure 8.8.

The key technique for detecting the COVID-19 disease is the CT scan, where the increase in the volume and density of CT scan images is positive evidence for confirmed COVID-19 case. The proposed framework allows the radiologists to efficiently decide the suspected cases which would otherwise take longer time if done manually by the radiologists [41].

Using deep learning, the COVID-19 infection can be predicted using X-ray images. If the epithelial cells of respiratory systems are affected by presence of COVID-19, the X-rays can be used to analyze this presence. The dataset of X-rays of normal people and X-rays of COVID-19 people can be used to train the model, and with the use of CNN and deep learning, the model will be able to detect the presence of COVID-19 virus in X-rays [42]. Figure 8.9 shows the X-ray dataset of normal and COVID-19 positive people and Figure 8.10 shows detection of COVID-19 using algorithm.

Use of medical imaging and the analysis of medical image data are growing very rapidly in the medical field. Due to advancement in the imaging technologies, large amount of data is available with medical applications being increasingly developed to use the data fruitfully and better methods of data analysis and algorithms are in demand. The operation of minimal intervention is one such example. Due to the advancement of methods in real-time imaging, major surgeries are being made possible which also require the surgical tools to have precise and automatic tracking. The methodologies in Deep Learning help in the analysis of medical image data by significant framework in the medical field [43].

Brain Tumor Segmentation is one of the applications of computer vision-based Deep Learning. In the United States, more than 20,000 people are diagnosed with tumor of spinal cord and brain of primary stage every year. The image processing of the brain tumor can be used to detect the tumor fall extensions where the anatomical structure is being segmented. The anatomical structures having unexpected shapes such as tumors in soft tissues can be challenging for automatic segmentation and the supervision of humans is required for complete segmentation [44]. Figure 8.11 shows the segmentation of the tumor in the brain. Active contours can be employed for tumor delineating with the aid of probabilistic maps. The successful segmentation of the brain tumor comes from the convergence of active contours at level set.

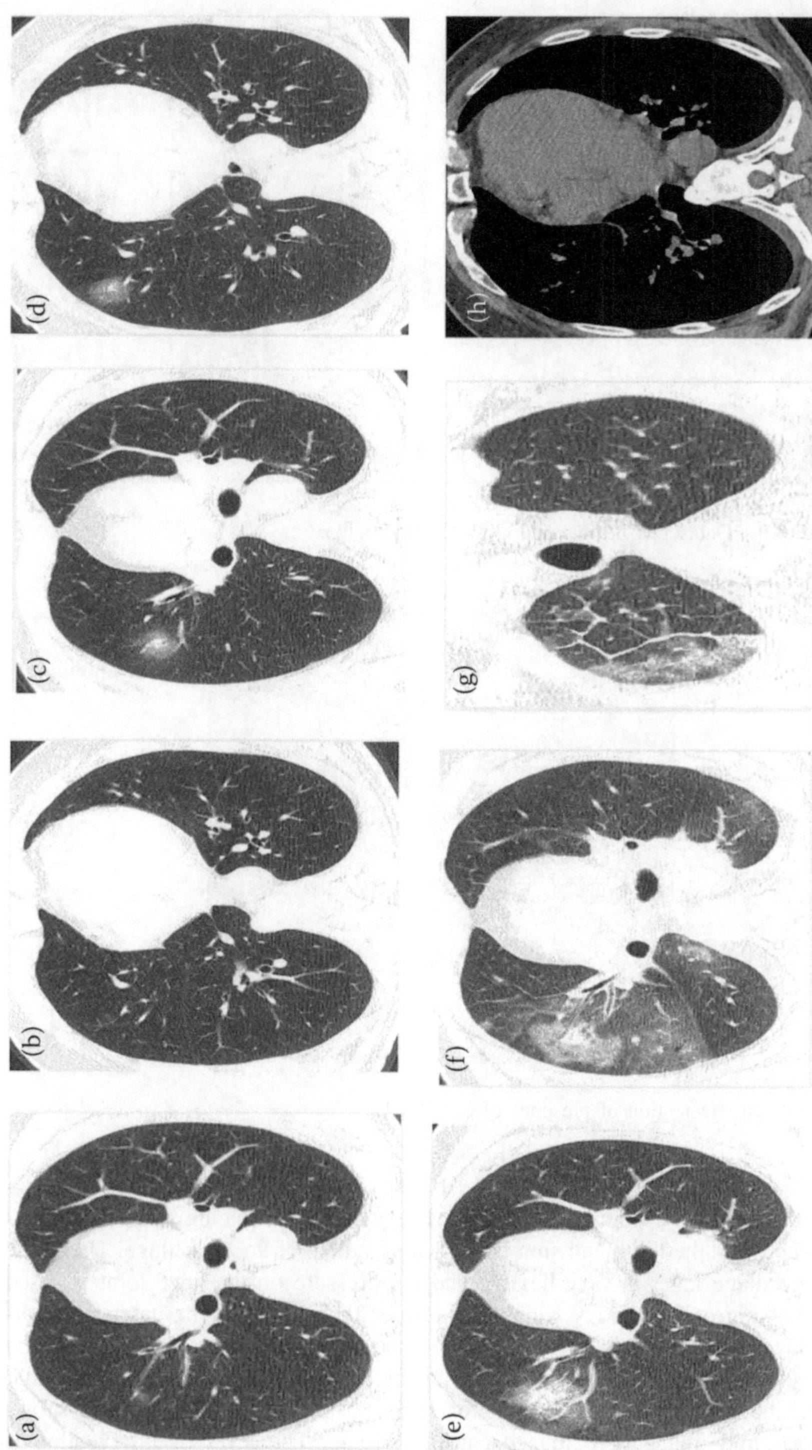

FIGURE 8.8 CT scan images of a suspected COVID-19 case.

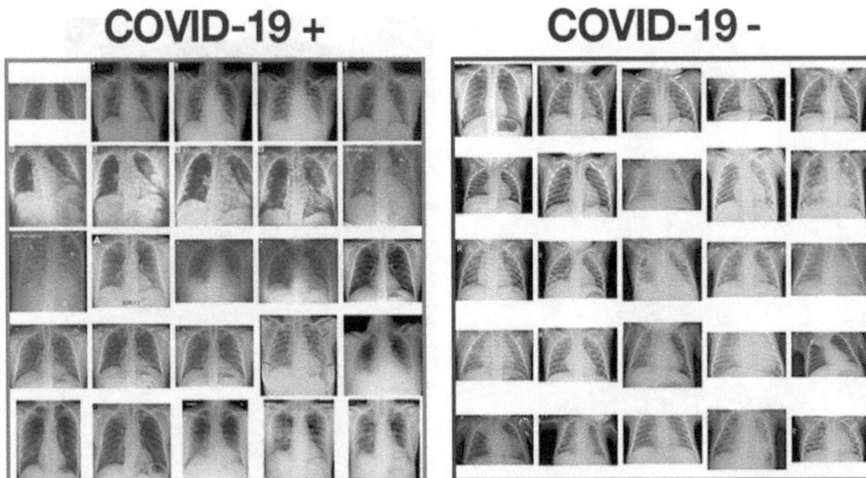

FIGURE 8.9 Dataset of normal and COVID-19 people.

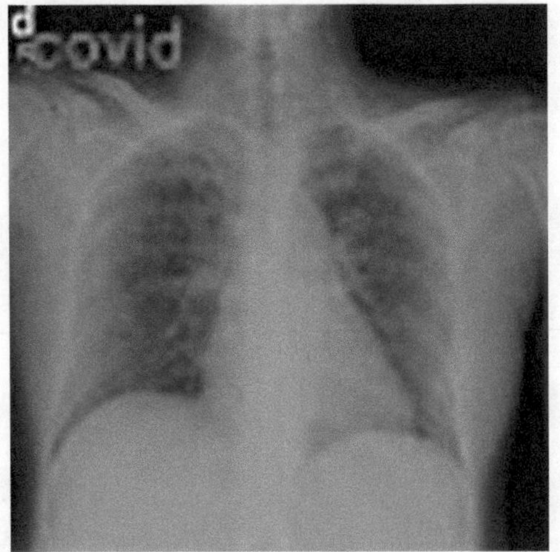

FIGURE 8.10 Detection of presence of COVID-19.

Another application of the computer-based vision in deep learning is the measuring the density of damaged cartilage. The shock impulses are damped and absorbed between the thigh and shin bones with the help of knee cartilage. The tissue of the cartilage layer is very flexible and the pressure on the knee joint is eased when walking or from body weight because of the cartilage. The damage to cartilage layer needs a clinical intervention such as encouraging new cartilage by drilling small holes or by totally replacing the knee cartilage. The ultrasonic imaging can be used to observe the state of cartilage erosion [45]. Figure 8.12 shows the ultrasound image of damaged cartilage tissue. The level of intervention is

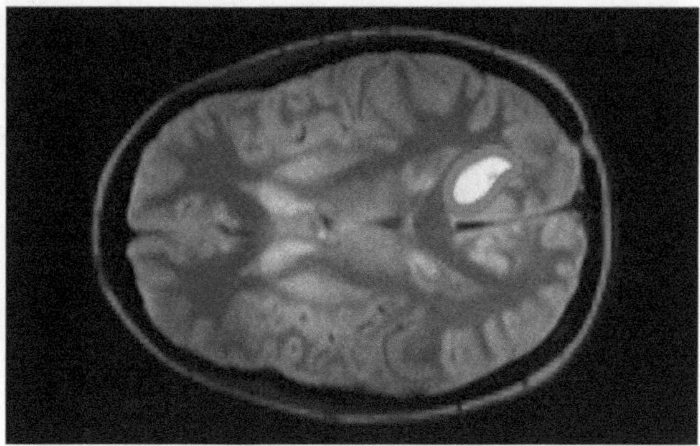

FIGURE 8.11 Segmentation of the brain tumor.

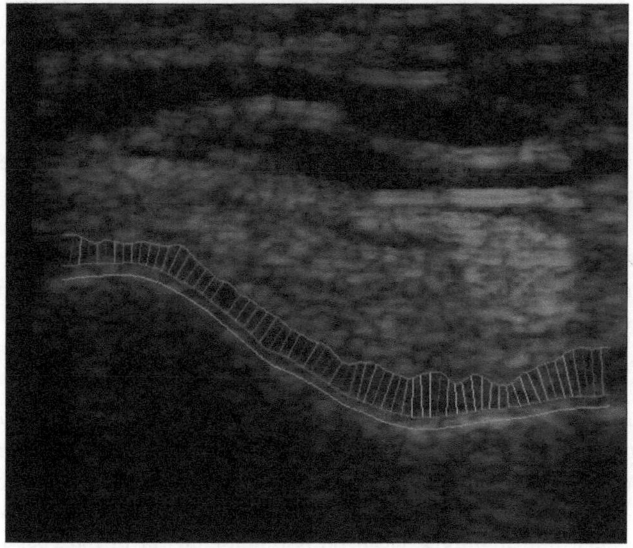

FIGURE 8.12 Ultrasound image of damaged cartilage tissue.

evaluated by trained orthopedics and the physicians access the knee cartilage thickness. The computational tools measure and analyze the output images.

Segmentation of skeletons and bones is another application of computer vision-based deep learning. The image of the bone can be easily obtained by Computed Tomography (CT) scan. Figure 8.13 shows the CT scan of the bones and skeleton. The fractured bones with the bone tissue easily identified through simple thresholding can be observed by 3-D modeling which has become necessary and significant in medical applications. The bone density can be assessed as a measurement in the CT scan which is the actual rate of intensity. Manual segmenting had been

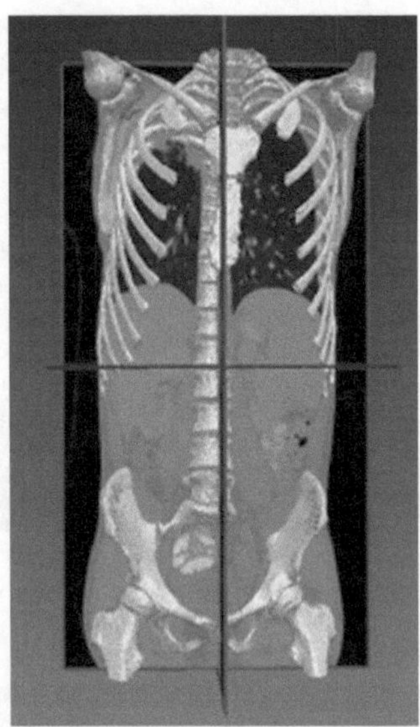

FIGURE 8.13 CT scan of bones and skeleton.

used for bone geometry segmentation but it is a very long process and error-prone. Deep learning models combined with the computer vision technology are able to provide the algorithm that segments the bones accurately in CT scans with robustness and high speed [46].

The tumor cells can be automatically segmented through machine learning methods. Examining the tumor cells manually and visually is highly time consuming and, in case of rapid intervention, this method is not readily available. This makes the manual segmentation a very unpractical task by the experts. Therefore, the algorithms based on computer vision and machine learning are proposed in which the tumor cells could be automatically segmented. The quantification tasks can be autonomously performed by these algorithms which scan and analyze the histological tissue at a very fast pace. Precious money and time can be saved by incorporating learning methods based on computer vision in the system of tumor cell segmentation [43].

8.3.6 IoT-Based System as Counter Measure for Bioweapon Against Crop War

In the threats described in previous section, smart farming, IoT based sensors (humidity, soil moisture, light, temperature, pesticide detection, etc.) are used to combat attack from bioweapons. Further, these are also used to increase the crop production and weather forecast based automatic irrigation system are used which optimizes water usage and eliminates wastage of water [47]. Figure 8.14 shows various applications of IoT in agriculture industry. During bio attacks on crops, the pre-symptoms can be detected at an early stage and proper precautions and

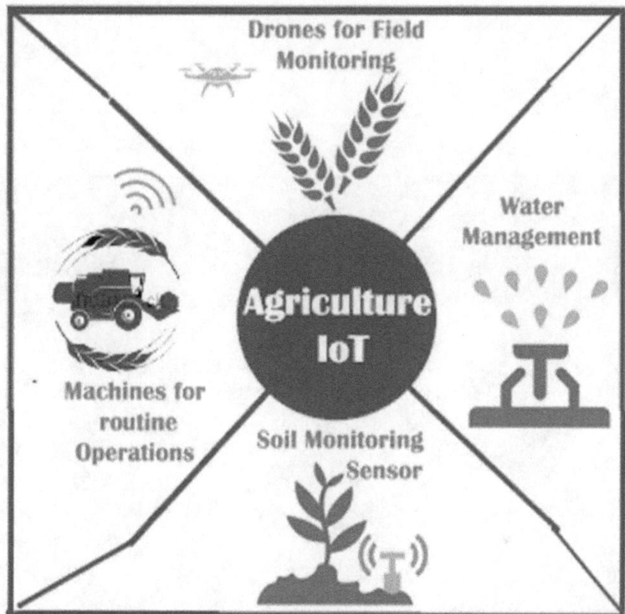

FIGURE 8.14 Applications of IoT-based smart agriculture devices.

fertilizers can be provided to plants. Several applications have been developed though computer vision and deep learning which can detect the type of disease of plants and to cure the suggestions are also provided. Figure 8.15 shows the data set for deep learning. The algorithm using deep learning and image processing is able to recognize infected leaf and stem, measure the affected area, it can find the shape of infected region, determine the color of infected region and can also influence the shape and size of crop [48,49]. The farmer can click the photograph of the crop and can upload it to the system via desktop application or mobile app. Using artificial intelligence and deep learning, and using the uploaded image, the disease can be detected promptly.

8.4 CONCLUSION

Other than natural disasters, we face threats of bioweapons and bioterrorism with genetically engineered agents. The bioweapon with a delivery system which can inject, spray or throw projectile of genetically engineered genes to infect humans and crops are increasing day by day. In this chapter, various genetically engineered pathogens, and the effect of biological agents are discussed. To counter the effects which is spread by these bio agents, several computer-based techniques are used. To provide faster, accurate results with computer vision and deep learning, the algorithms are helping to detect anomlies in the medical area in cost-effective and timely ways. Artificial intelligence, big data, computer vision, computer-assisted systems and IoT-based systems are providing solutions in any pandemic. Computer-

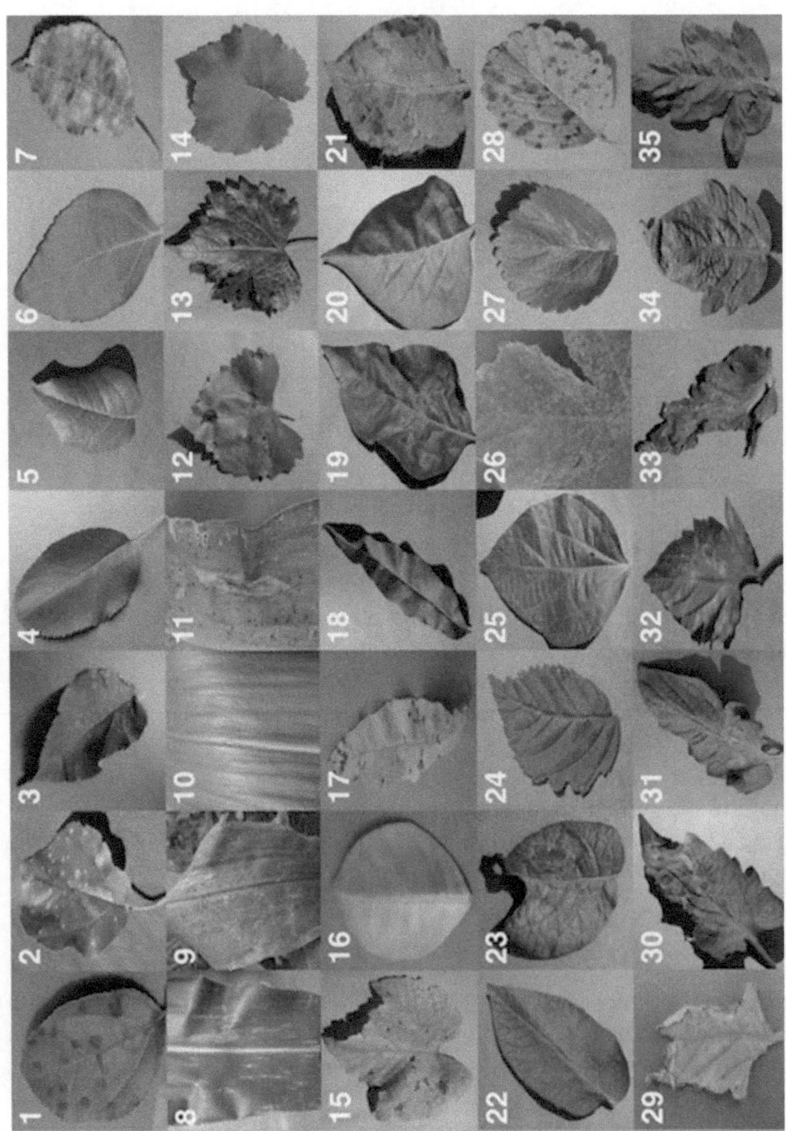

FIGURE 8.15 Image set for deep learning.

based surgery is helping doctors and, now, even the doctors are able to do surgery at remote locations. IoT-based smart farming devices are not only used to increase the crop production; also, it can monitor the crop damage done from any biological agent during crop war while suggesting the cure to the farmer. With the increase in bio war technology, the counter measures using computer-assisted techniques are also growing.

REFERENCES

[1]. Hassani, Morad, Mahesh C. Patel, and Liise-anne Pirofski. "Vaccines for the prevention of diseases caused by potential bioweapons." Clinical Immunology 111, no. 1, 1–15, 2001.

[2]. Murch, Randall S. "Forensic perspective on bioterrorism and the proliferation of bioweapons." Firepower in the lab: Automation in the fight against infectious diseases and bioterrorism , pp. 211–213. Joseph Henry Press, United State of America, 2001.

[3]. Ainscough, Michael. "Next generation bioweapons: Genetic engineering and BW." The gathering biological warfare storm, pp. 269–270. The Homeland Security Digital Library, United State of America, 2004.

[4]. Rumyantsev, Sergey N. "The best defence against bioweapons has already been invented by evolution." Infection, Genetics and Evolution 4, no. 2, 159–166, 2004.

[5]. Boddie, Crystal, Matthew Watson, Gary Ackerman, and Gigi Kwik Gronvall. "Assessing the bioweapons threat." Science 349, no. 6250, 792–793, 2015.

[6]. Tucker, Jonathan B. "The current bioweapons threat." In Bio preparedness and public health, pp. 7–16. Springer, Dordrecht, 2013.

[7]. Myint, Steven H. "Human coronavirus infections." In the Coronaviridae, pp. 389–401. Springer, Boston, MA, 1995.

[8]. Novel Swine-Origin Influenza A (H1N1) Virus Investigation Team. "Emergence of a novel swine-origin influenza A (H1N1) virus in humans." New England Journal of Medicine 360, no. 25, 2605–2615, 2009.

[9]. Writing Committee of the WHO Consultation on Clinical Aspects of Pandemic (H1N1) 2009 Influenza. "Clinical aspects of pandemic 2009 influenza A (H1N1) virus infection." New England Journal of Medicine 362, no. 18, 1708–1719, 2010.

[10]. Martina, Byron E. E., Bart L. Haagmans, Thijs Kuiken, Ron A. M. Fouchier, Guus F. Rimmelzwaan, Geert Van Amerongen, J. S. Malik Peiris, Wilina Lim, and Albert D. M. E. Osterhaus. "SARS virus infection of cats and ferrets." Nature 425, no. 6961, 915, 2003.

[11]. Stadler, Konrad, Vega Masignani, Markus Eickmann, Stephan Becker, Sergio Abrignani, Hans-Dieter Klenk, and Rino Rappuoli. "SARS—beginning to understand a new virus." Nature Reviews Microbiology 1, no. 3, 209–218, 2003.

[12]. Haagmans, Bart L., Judith M. A. van den Brand, V. Stalin Raj, Asisa Volz, Peter Wohlsein, Saskia L. Smits, Debby Schipper et al. "An orthopoxvirus-based vaccine reduces virus excretion after MERS-CoV infection in dromedary camels." Science 351, no. 6268, 77–81, 2016.

[13]. Tang, Xian-Chun, Sudhakar S. Agnihothram, Yongjun Jiao, Jeremy Stanhope, Rachel L. Graham, Eric C. Peterson, Yuval Avnir et al. "Identification of human neutralizing antibodies against MERS-CoV and their role in virus adaptive evolution." Proceedings of the National Academy of Sciences 111, no. 19, E2018–E2026, 2014.

[14]. Alper, J. "From the bioweapon's trenches, new tools for battling microbes." Science 284, 1754–1755, 1999.
[15]. Atlas, R. M. "Biological weapons pose challenge for microbiological community." ASM News 64, 383–388, 1998.
[16]. Cole, C. A. "The spectre of biological weapons." Scientific American 275, 60–65, 1996.
[17]. Steidler, Lothar. "Genetically engineered probiotics." Best Practice & Research Clinical Gastroenterology 17, no. 5, 861–876, 2003.
[18]. Center, US Air Force Counterproliferation. "Genetically Engineered Pathogens."
[19]. Nida, T. K., and J. K. Maham. "Bioweapons-Future of warfare." Research & Reviews : Research Journal of Biology (RRJOB) 5, no. 4, 16–18, 2017.
[20]. Aldhous, Peter. "Biologists urged to address risk of data aiding bioweapon design." Nature 414, 237–238, 2001.
[21]. Martin, John. "Severe stealth virus encephalopathy following chronic-fatigue-syndrome-like illness: Clinical and histopathological features." Pathobiology 64, no. 1, 1–8, 1996.
[22]. Bested, A. C., and L. M. Marshall. "Review of myalgic encephalomyelitis/chronic fatigue syndrome: An evidence-based approach to diagnosis and management by clinicians." Reviews on Environmental Health 30, 223–249, 2015.
[23]. Sackstein, Robert. "A revision of Billingham's tenets: The central role of lymphocyte migration in acute graft-versus-host disease." Biology of Blood and Marrow Transplantation 12, no. 1, 2–8, 2006.
[24]. Zhao, Xiaojun, and Shuguang Zhang. "Molecular designer self-assembling peptides." Chemical Society Reviews 35, no. 11, 1105–1110, 2006.
[25]. Roopashree, Anitha et al. "Machine learning approach: Enriching the knowledge of Ayurveda from Indian medicinal herbs." Challenges and applications of data analytics in social perspectives. IGI Global Publications, United State of America, 2020.
[26]. Hosny, Ahmed, Chintan Parmar, John Quackenbush, Lawrence H. Schwartz, and Hugo J. W. L. Aerts. "Artificial intelligence in radiology." Nature Reviews Cancer 18, no. 8, 500–510, 2018.
[27]. Kim, Myung K. "Digital holographic microscopy." Digital holographic microscopy, pp. 149–190. Springer, New York, NY, 2011.
[28]. Adams, Ludwig, Werner Krybus, Dietrich Meyer-Ebrecht, Rainer Rueger, Joachim M. Gilsbach, Ralph Moesges, and Georg Schloendorff. "Computer-assisted surgery." IEEE Computer Graphics and Applications 10, no. 3, 43–51, 1990.
[29]. Wang, Yulun, Modjtaba Ghodoussi, Darrin Uecker, James Wright, and Mangaser Amant. "Modularity system for computer assisted surgery." U.S. Patent 6,728,599, issued April 27, 2004.
[30]. DiGioia III, Anthony M., David A. Simon, Branislav Jaramaz, Michael K. Blackwell, Frederick M. Morgan, Robert V. O'Toole, and Takeo Kanade. "Computer-assisted surgery planner and intra-operative guidance system." U.S. Patent 6,205,411, issued March 20, 2001.
[31]. Kienzle III, Thomas C. "Enhanced graphic features for computer assisted surgery system." U.S. Patent 6,917,827, issued July 12, 2005.
[32]. Wang, Yulun, Modjtaba Ghodoussi, Darrin Uecker, James Wright, and Amante Mangaser. "Modularity system for computer assisted surgery." U.S. Patent 6,892,112, issued May 10, 2005.
[33]. Bates, David W., Suchi Saria, Lucila Ohno-Machado, Anand Shah, and Gabriel Escobar. "Big data in health care: Using analytics to identify and manage high-risk and high-cost patients." Health Affairs 33, no. 7, 1123–1131, 2014.

[34]. Kaur, Sanam Preet, and Manvinder Sharma. "Radially optimized zone-divided energy-aware wireless sensor networks (WSN) protocol using BA (bat algorithm)." IETE Journal of Research 61, no. 2, 170–179, 2015.

[35]. Andreu-Perez, Javier, Carmen C. Y. Poon, Robert D. Merrifield, Stephen T. C. Wong, and Guang-Zhong Yang. "Big data for health." IEEE Journal of Biomedical and Health Informatics 19, no. 4, 1193–1208, 2015.

[36]. Suma, M. R. "Hybrid cloud - Intra domain data security and to address the issues of interoperability." International Journal of Recent Technology and Engineering 8, no. 15, 340–344, 2019.

[37]. Sharma, Manvinder, and Harjinder Singh. "SIW based leaky wave antenna with semi C-shaped slots and its modelling, design and parametric considerations for different materials of dielectric." In 2018 Fifth International Conference on Parallel, Distributed and Grid Computing (PDGC), pp. 252–258. IEEE, 2018.

[38]. Nachtigall, I., S. Tafelski, M. Deja, E. Halle, M. C. Grebe, A. Tamarkin, A. Rothbart et al. "Long-term effect of computer-assisted decision support for antibiotic treatment in critically ill patients: A prospective 'before/after' cohort study." BMJ Open 4, no. 12, e005370, 2014.

[39]. Gil, Miguel, Pedro Pinto, Alexandra S. Simões, Pedro Póvoa, M. M. da Silva, and L. Lapao. "Co-design of a computer-assisted medical decision support system to manage antibiotic prescription in an ICU ward." Studies in health technology and informatics, p. 228. IOS Press, United State of America, 2016.

[40]. Wahabi, Hayfaa Abdelmageed, Samia Ahmed Esmaeil, Khawater Hassan Bahkali, Maher Abdelraheim Titi, Yasser Sami Amer, Amel Ahmed Fayed, Amr Jamal et al. "Medical doctors' offline computer-assisted digital education: Systematic review by the digital health education collaboration." Journal of Medical Internet Research 21, no. 3, e12998, 2019.

[41]. Li, Lin, Lixin Qin, Zeguo Xu, Youbing Yin, Xin Wang, Bin Kong, Junjie Bai et al. "Artificial intelligence distinguishes covid-19 from community acquired pneumonia on chest CT." Radiology 200905, 2020.

[42]. Gupta, Anuj Kumar, Manvinder Sharma, Ankit Sharma, and Vikas Menon. "A study on SARS-CoV-2 (COVID-19) and machine learning based approach to detect COVID-19 through X-ray images." International Journal of Image and Graphics 2140010, 2020.

[43]. Isitha Banerjee. "Brain tumor image segmentation and classification using SVM, CLAHE and ARKFCM." Intelligent decision support systems, applications in signal processing, pp. 53–70. De-Gruyter, Berlin, 2019.

[44]. Zhao, Xiaomei, Yihong Wu, Guidong Song, Zhenye Li, Yazhuo Zhang, and Yong Fan. "A deep learning model integrating FCNNs and CRFs for brain tumor segmentation." Medical Image Analysis 43, 98–111, 2018.

[45]. Kaur, Navpreet, and Manvinder Sharma. "Brain tumor detection using self-adaptive K-means clustering." In 2017 International Conference on Energy, Communication, Data Analytics and Soft Computing (ICECDS), pp. 1861–1865. IEEE, 2017.

[46]. Gjertsson, Konrad, Kerstin Johnsson, Jens Richter, Karl Sjöstrand, Lars Edenbrandt, and Aseem Anand. "A novel automated deep learning algorithm for segmentation of the skeleton in low-dose CT for [(18) F] DCFPyL PET/CT hybrid imaging in patients with metastatic prostate cancer". Journal of Clinical Oncology 25, no. 25, Wolters Kluwer Health, Philadelphia, 2020.

[47]. Reddy, G. Balakrishna, and K. Ratna Kumar. "Quality improvement in organic food supply chain using blockchain technology." Innovative product design and intelligent manufacturing systems, pp. 887–896. Springer, Singapore, 2020.

[48]. Sharma, Manvinder, Bikramjit Sharma, Anuj Kumar Gupta, and Bhim Sain Singla. "Design of 7 GHz microstrip patch antenna for satellite IoT- and IoE-based devices." In The International Conference on Recent Innovations in Computing, pp. 627–637. Springer, Singapore, 2020.
[49]. Elgabry, Mariam, Darren Nesbeth, and Shane D. Johnson. "A systematic review protocols for crime trends facilitated by synthetic biology." Systematic Reviews 9, no. 1, 22, 2020.

9 An Automated Hybrid Transfer Learning System for Detection and Segmentation of Tumor in MRI Brain Images with UNet and VGG-19 Network

S. Sandhya[1], *Dr. M. Senthil Kumar*[2], *and Dr. B. Chidhambararajan*[3]

[1]Research Scholar, Department of Information Technology, SRM Valliammai Engineering College

[2]Associate Professor, Department of Computer Science and Engineering, SRM Valliammai Engineering College

[3]Professor and Principal, SRM Valliammai Engineering College

CONTENTS

9.1 INTRODUCTION

MRI – Magnetic Resonance Imaging – is one of the well-recognized modalities in medical imaging applications as it visualizes various abnormalities in the brain by exhibiting the soft tissue details. By using the information captured from MRI, the

DOI: 10.1201/9781003194415-9

computer aided diagnosis system is helping the medical experts in diagnosing the abnormalities in the brain and to carry out a treatment accordingly. Brain tumor is a condition where the growth of brain cells or tissues is abnormal and where brain cells grow in an uncontrolled manner. Based on the scale of malignant cells, brain tumor can be categorized as low grade glioma, which in turn includes grade 1 and 2; and high grade glioma, with grade 3 and 4. By diagnosing the nature of the tumor like grade, size of the malignant cell and the type, a treatment will be decided by the medical experts. MRI exhibits contrast of brain tissues more effectively than can be done by the modalities like computed tomography. Various MRI modalities namely T1, T2, FLAIR, T1-CE are used for brain tumor detection.

Malignant brain tissues in analogous to the normal brain cells are identified by segmenting the brain tumor, and with the results of segmented brain tissues, treatment can be planned, given planning is helpful in tracking the advancement of the tumor cells. Physical segmentation is an intense task in terms of time and complexity of the task as it needs to handle the heterogeneous characteristics of the tumor cells. The limitation in segmenting the brain tumor manually is in the emergence of the need to carry out the segmentation of brain tumor automatically. There are two types of brain tumor segmentation methods [1]. The first is the generative model in which information about the manifestation and location of the normal-healthy and malignant brain tissue needs to be provided prior to segmentation being carried out. Example for generative model is atlas-based models. The second type, i.e. discriminative model like machine learning based SVM and random-forests, relies upon prior information to a small scale but where the image features are learnt from the annotated trained images to a large scale.

In the realm of Machine Learning (ML) approaches, techniques like convolutional neural networks become more dominantly prevailing in the field of deep learning. A variation of CNN which is known as deep convolutional neural network automatically extracts the significant features of the images by tuning convolutional and pooling layers of the networks. Due to its prevailing nature, CNN is used in a wide range of medical applications. Depending upon the size of training data being utilized, potential of CNN is decided. CNN starts to over-fit if the size of the data is undersized. In this case, classification's efficacy is improved by utilizing the techniques to carry out transfer learning.

Tasks like segmentation and classification are performed in an excellent fashion by feeding the network with multi-convolutional layers to generate better results, rather than the manual segmentation and classification using convolutional neural network. In recent times, diverse CNN models are being used for segmentation of medical images like RestNet, SgeNet and GoogLeNet.

9.2 RELATED WORKS

Multimodal MRI is the primary and the most significant method for viewing and diagnosing the tumor. Once diagnosed, it is critical to formulate the diagnosis and chalk out the treatment plan as features like location, scale of growth and quantity of the tumor cells need to be determined for the accurate segmentation. With the advancement in Computer Vision, it is possible to perform the segmentation task automatically. In specific cases, convolutional neural networks perform in a better way in automating the tumor segmentation. With the success of CNN, tasks like segmentation and classification are carried out with ease because of which the medical experts show interest in adopting those CNN models for processing the various medical imaging-related applications, thus, providing high throughput and accuracy.

In medical image processing, exercising Deep Neural Networks, at first, used CNN based architecture to carry out the task of pixel-wise classification of electron-microscopy neuron images into membrane and non-membrane pixels. But, there are several challenges in medical image analysis and segmentation as segmentation exhibits heterogeneity in representing the patient data, given the same pathology can exist across patients in various manners. In performing segmentation of medical images, the data available is a major constraint and to be considered before deciding segmentation method to apply. For instance, when size of data is small and even where the available data is inconsistent, then segmentation to be done with CNN trained on such data is to invite the problem of over-fitting. Irrespective of these shortcomings, CNN based techniques perform better in terms of carrying out the task of segmenting the medical images with greater accuracy compared with the generative models [2].

[3] used CNN for MRI brain tumor segmentation, in which the proposed system involves two-phase training in order to handle the tumor label imbalance by feeding the network with features inclusive of local level and global level. [4] anticipated 3D CNN model for the segmentation of tumor with 11 layers of learning, namely Deep Medic. Pereira *et al.* implemented the brain tumor segmentation of LGG and HGG using two CNN. [5] implemented UNet for segmenting the neuron structures from the electron microscopic images.

[6] proposed fully convolutional architecture where the input to the network is a complete image and the volume of semantic segmentation is the result. [7] explored the fully convolutional architecture by using VGG model and demonstrated that the proposed system provides more accuracy when compared to pixel-based techniques with low cost of computation. [8] proposed a system which performed the segmentation of multimodal MRI brain images at the volumetric level. [3] segmented the individual MRI slices with which the process of segmentation is taken forward towards examining the presence of brain tumor. [9] proposed segmentation of tumor in MRI brain by combining CNN with other numerical methods.

[10] implemented pixel-based UNet for performing the segmentation of brain tumor in MRI images. [11] achieved segmentation of brain tumor by utilizing fully convolutional networks.

Towards the face of achieving better performance at a higher level during the processes such as detecting and segmenting tumor in the brain tissues, where the data available is minimal, transfer learning techniques are being used widely. [12] demonstrated that CNN learn analogous image features from the trained datasets even when the networks differ in terms of tasks and statistics. Despite the benefit in using transfer learning in analyzing the medical images, however, the availability of datasets for medical images is, in turn, a critical factor. The biomedical image datasets greatly vary in size, i.e. large-scale datasets of medical images are rare to find and it is difficult to obtain such datasets. Also the nature of datasets tends to depend upon the patients and tasks which make the transfer learning more intricate. Noh *et al.* proposed full convolutional neural networks for full image segmentation by using VGG-16 model [7].

[13] applied FCN to alleviate the process of training the data and to achieve better prediction performance by using the encoders on the ImageNet dataset.

A fully convolutional network uses only the convolutional layers and it segments the whole image at once instead of processing one pixel at a time by incurring more cost for its high memory necessities. This led to the design of UNet model which is utilized specifically for the purpose of segmenting images in the field of medical imaging [9].

[14–16] used convolutional neural network, and the concepts of deep learning was involved for performing segmentation. [17,18] and [19] proposed a fully convolutional network as it yields greater performance in semantic segmentation.

[20] proposed an automatic brain tumor segmentation system by deploying the fully convolutional network, along with VGG-16 network, thus implementing the transfer learning.

[21] proposed a super-pixel mechanism for performing brain tumor detection and segmentation that is based on transfer learning to detect the LGG, HGG and normal as three categories from the brain MRI images. It utilized VGG-19 model to perform transfer learning for detecting the brain tumor.

[22] proposed deep learning techniques by combining the CNN that is relying on UNet, along with the base of transfer learning by adopting the VGG-16 for grading the brain tumor. The proposed system designed was able to identify the LGG with better sensitivity, accuracy and specificity. [2] proposed 3 architectures that are based on CNN for the brain tumor segmentation of MRI images from multi-modalities. [23] deployed FCN by integrating the UNet with VGG-16 model for detecting the ROI and non-ROI of the brain MRI images.

[24] achieved transfer learning by incorporating deep learning architecture in performing the brain tumor detection from 2-dimensional MRI image slices and achieved 98% of accuracy in classification. [25] proposed detection of brain tumor by using deep learning techniques, thus, achieving better results in terms of specificity and sensitivity.

9.3 PROPOSED SYSTEM

Image segmentation has achieved significant improvement, in terms of its results, by utilizing fully convolutional network (FCN) with the integration of transfer learning. Various deep learning techniques like GoogLeNet, RestNet and SegNet had been deployed to perform the task of segmentation depending on the nature of application [20]. A fully convolutional network is one of the techniques to perform semantic segmentation under the deep learning approaches. Fully convolutional network can be recognized as an advance technique over the existing CNN. A typical CNN comprises of convolution, pooling and finally fully connected layers as its essential components. Here, in the FCN, the fully connected layer has been replaced by convolutional layer with which the FCN is capable of classifying each and every pixel from the input images thus providing the ability to the network to predict the output over the subjective input. Three layers namely Height (h), Width (w) and Depth (d) are identified from the input images. Spatial dimension is depicted by the layer's h and w, whereas, the color channel is represented using the layer d (when d = 1, the image possesses the gray-scale intensity; and it is RGB, when d = 3). If the input date vector x_{ij} exists in a location (i,j) of a specific layer, then the output vector y_{ij} is calculated by using the formula given in Equation (9.1).

$$y_{ij} = f_{ks\ (\{x_{si}+\delta_{i,s}+\delta_j\},0\leq\delta_i,\ \delta_j\leq k)} \tag{9.1}$$

In Equation (9.1), size of the kernel is given as k, factor of sub-sampling is represented as s, f_{ks} represents the nature of the layer used.

For the image segmentation of the biomedical images, UNet architecture is the most widely-used fully convolutional network. The name of the architecture is implied due to its U shape as it holds two ways. The left side represents the layer for contracting, i.e. encoder, and the right side of the U shape represents the layer for expanding, i.e. decoder. The output is a feature map/vector in the encoder holding the information from the input, whereas, the decoder is the same as encoder but it acts in the opposite fashion where it extracts the feature map from the expanding layer, thus, providing the exact match as output for the input. The number of feature maps is reduced by increasing the input matrix, which is the process carried out by the encoder. The process involved in the decoder is to minimize the number of feature maps by restoring the matrix to its actual size. Thus, the segmentation results are figured out by taking all the pixels of the image with the ground truth.

The UNet model is used to make decisions by transmitting the feature map from the subsequent levels of the contracting path, which is analogous to the path of expanding, which makes the classifier capable of handling the features at varying scales and complexities. Hence, UNet is capable of working on datasets that are relatively small in size. Often, more time is spent on execution in the UNet architecture. In order to overcome this shortcoming, transfer learning is integrated with UNet model. Transfer learning uses pre-trained models to advance the target task learning by incorporating the information from the source. Distinct architectural models like AlexNet, ZFNet, VGG Net and LeNet are considered as suitable models for developing the hybrid approach, along with UNet model. The performance of such hybrid models can be improved by using VGG network, where the size of kernel is small with a deep convolutional neural network.

The proposed scheme aims at developing a hybrid model, where the UNet is integrated with another model named VGG-19. The aim of this integration is to overcome the limitation of UNet, which takes considerable amount of time for execution by reducing the layers and the parameters of UNet model. The reason behind the selection of VGG-19 network is that it consists of the contracting layer similar to UNet, and the parameter count is less than that of UNet. Models of segmentation consist of contracting and expanding layer, in general. This is modified in the VGG-19 network by adding extra expanding layers that hold up-sampling layers, and, at the closing stages of VGG-19 network's convolutional layers. The procedure of adding additional layers is repeated until the network gets a symmetric nature and resembles the U shape. Hence, in the hybrid model, UNet and VGG-19 network model, the contracting layer will depict the VGG-19. UNet VGG-19 network with transfer learning is trained for the brain tumor MRI images.

The goal of the hybrid model is to reduce the time taken for execution and to fasten the time of training the dataset. The proposed architecture of the hybrid model with transfer learning is depicted in Figure 9.1.

The process of segmentation of brain tumor image along with transfer learning is depicted in Figure 9.2.

In order to achieve efficiency in performing the segmentation of medical images, the hybrid model of UNet with VGG-19 is used in the proposed system. The VGG-19 network is a kind of fully convolutional network which acts as a base and it is trained on ImageNet to carry out the classification of images. In drawing out the predominant outcomes of tumor segmentation, VGG-19 network is used with BraTs 2015 database. During the process of semantic segmentation, the pixel level labels and the images itself are fed to a pre-trained deep learning network in order to identify the fresh labels by training the network. The proposed hybrid network

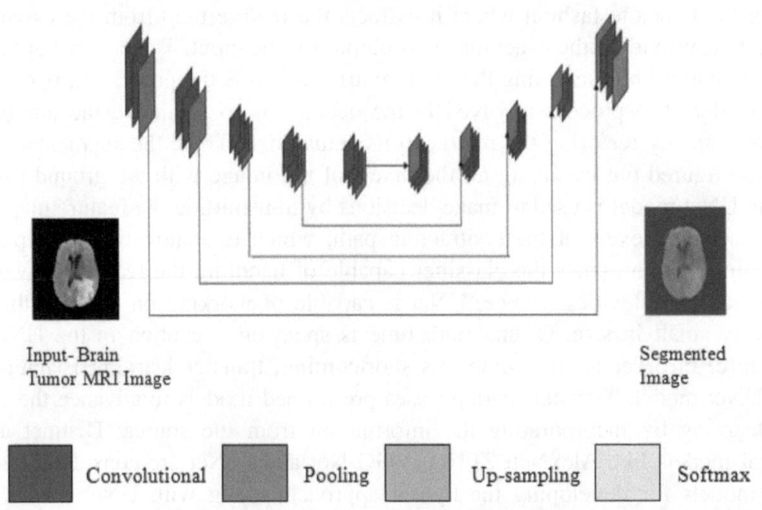

FIGURE 9.1 Proposed Hybrid Model Architecture with Transfer Learning.

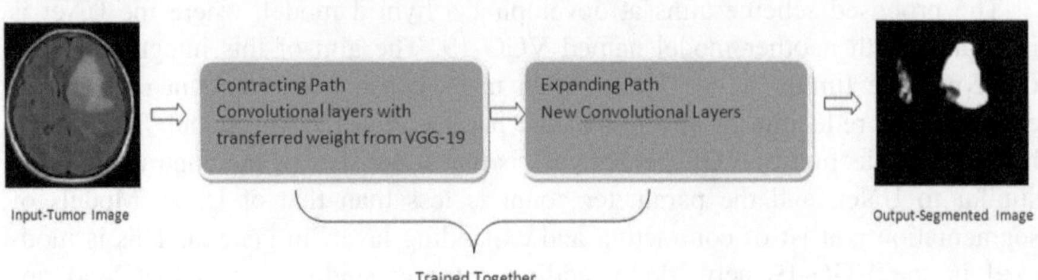

FIGURE 9.2 Image Segmentation with Transfer Learning.

comprises an encoder, decoder and, finally, a classification layer with which the images are classified at pixel level. In the process of detecting brain tumor, the proposed system uses VGG-19 network which is a pre-trained network based on transfer learning approach. The final layers of the network are replaced with new layers to detect the tumors under three categories such as Normal-0, LGG-1 and finally HGG-2.

Training options for the pre-trained network are specified as follows: batch size as 10 images per iteration at the mini level and number of epochs, where epoch represents the entire training cycle of training dataset with 2 to 8 epochs inclusive of step-size with value of 2; and, finally, the rate of learning is reduced to undersized value in order to slow down the learning carried out at the level of transferred layers. Hence, the proposed sculpt minimizes the amount of parameters and the computational time too.

9.4 EXPERIMENTAL SETUP AND RESULTS

The proposed work is evaluated with the BraTs 2015 database which aims at handling the challenge of tumor segmentation of multimodal brain images. Synthetic and clinical images of the glioma are both included in the database. The modalities like T1, T1-weighted and contrast enhanced, T2-weighted and, finally, FLAIR are captured, and annotated imageries from the man-made data are generated by the clinical experts to yield the truth data of opinion.

Figure 9.3 represents the FLAIR images of HGG with its annotated ground truth data.

The proposed framework uses the MRI FLAIR images of HGG as it best discriminates the malignant tissue. Analysis is also carried out at the pixel level to determine the relative amount of normal and the malignant brain tissue.

The investigational outcome for the proposed hybrid model is carried out on 64-bit operating system with processing speed of 3.4 GHz, intel core i7 processor and 16GB RAM using MATLAB. The proposed VGG-19 network in performing the transfer learning is evaluated up to epochs with count of 8, with its achieving maximum precision at the 6th epoch during the mean time of transit. Truthfulness

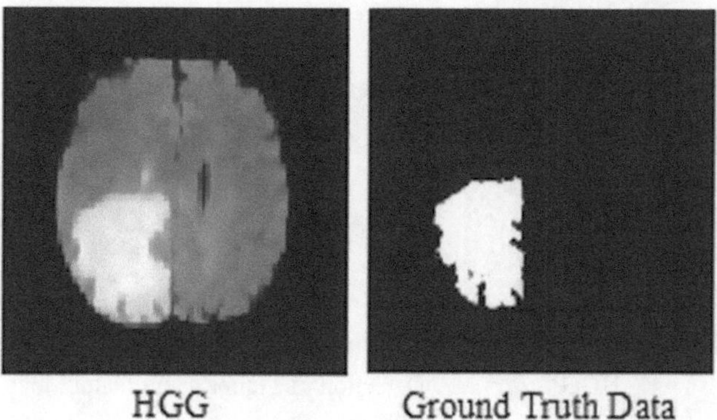

FIGURE 9.3 FLAIR MRI Image of HGG along with Its Ground Truth Data.

of classification is computed as a relative amount of exact computations to the total cumulated computations. In the field of medical imaging, accuracy is not alone sufficient to classify the patient's details as either normal or abnormal. Hence, measures like specificity, sensitivity and area under curve (AUC) are also to be computed. Specificity and sensitivity are relatively used to detect the normal from the tumor patients. AUC is used to measure the accuracy of specificity and sensitivity in terms of identifying the classes of normal and abnormal conditions.

The transfer learning model with VGG-19 achieves 99.8% of accuracy during training, 96.4% of accuracy during validation in the proportion of 7:3 of data during training and data during validation. Normal and abnormal classes of brain tumor are examined by exploring the network features in deep from the convolutional layers and as well as the deep-layers. While performing the detection of tumor in MRI brain images, complex features from the deepest layer has to be analyzed by generating the feature maps of the tumor cells with high accuracy. By integrating the features of the earlier layer's, the deepest convolutional layers can be constructed; thus, the features to detect the tumor area are obtained using the activation channel. Further, the feature map is generated by merging the strong activation channel along with the original image. Feature maps are used to highlight the tumor portions detected with the help of CNN model with transfer learning approach. Figure 9.4 provides details of original image and its feature map highlighting the tumor portion.

Mean dice index D is calculated as given in Equation (9.2) to assess effective segmentation.

$$\text{Dice Index}(D) = \frac{2\,TP}{2TP + FP + FN} \tag{9.2}$$

In Equation (9.2), True-Positive, False-Positive and False-Negative are respectively represented as TP, FP and FN. If value of D is 1, then it can be concluded the segmented tumor image and the annotated ground truth data are matching perfectly.

The proposed hybrid network model for tumor detection and segmentation is compared with the available classic techniques. Tables 9.1 and 9.2 exhibit proposed

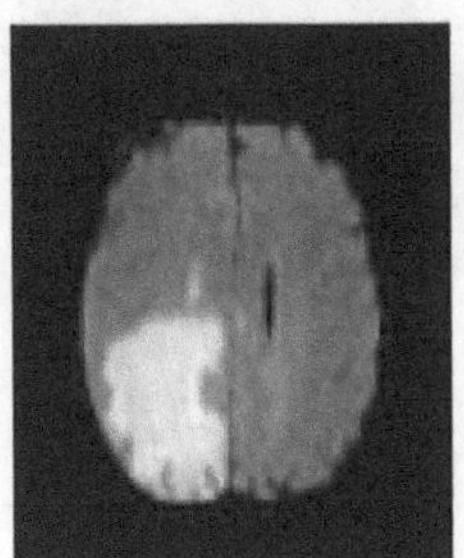

HGG

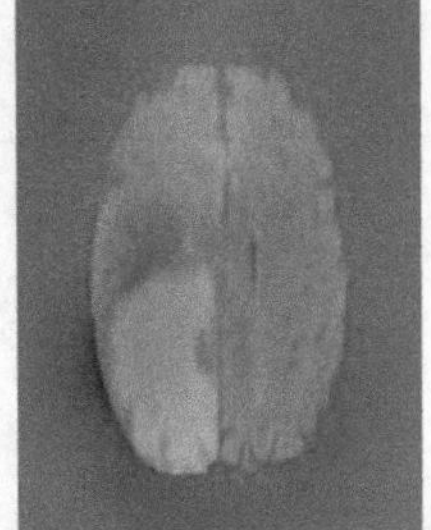

Detection of Tumor using feature map

FIGURE 9.4 Analysis of Tumor Detection at the Deepest Layers.

TABLE 9.1
Comparison of the Measure of Accuracy

Task	Techniques	Accuracy
Tumor detection	Random Forest	0.91
	Proposed	0.93
Tumor segmentation	CNN	0.82
	Proposed	0.99
Tumor grading	CNN	0.77
	Proposed	0.88

TABLE 9.2
Dice Index Comparison

Task	Techniques	Accuracy
Tumor detection	Random Forest	0.91
	Proposed	0.93
Tumor segmentation	CNN	0.82
	Proposed	0.84

model's efficacy over the techniques which are already available, in terms of accuracy as well as dice index.

Figures 9.5 and 9.6 pictorially depict evaluation results of the proposed system with that of the existing techniques.

9.5 DISCUSSION

The proposed model performs the brain tumor detection, segmentation and grading by involving the deep learning approaches that used CNN along with UNet, and the techniques of transfer learning by using pre-trained convolutional based VGG-19 network. During the first stage of the network, the input images of MRI brain are sliced such that it can be categorized as normal tissue, low grade and high grade of the condition glioma. During epoch 6, the hybrid network model achieves 99.8% of accuracy during training, and 96.4% of accuracy during the validation of data. During the testing phase, 99.3% of accuracy, 100% of specificity, 97.8% of sensitivity and 0.99 Area Under Curve are obtained. The average dice index value obtained by the proposed system is 0.93 for tumor detection, and 0.84 for tumor segmentation.

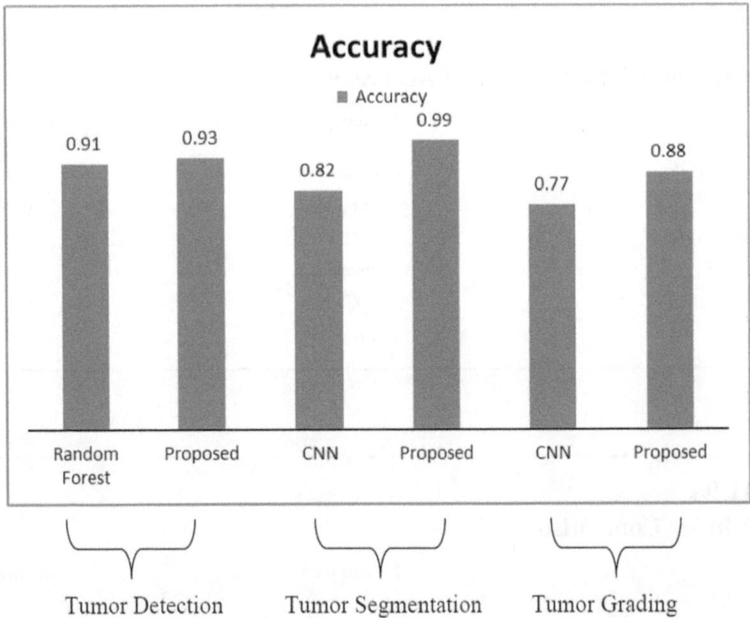

FIGURE 9.5 Performance Results of Proposed System in Terms of Accuracy.

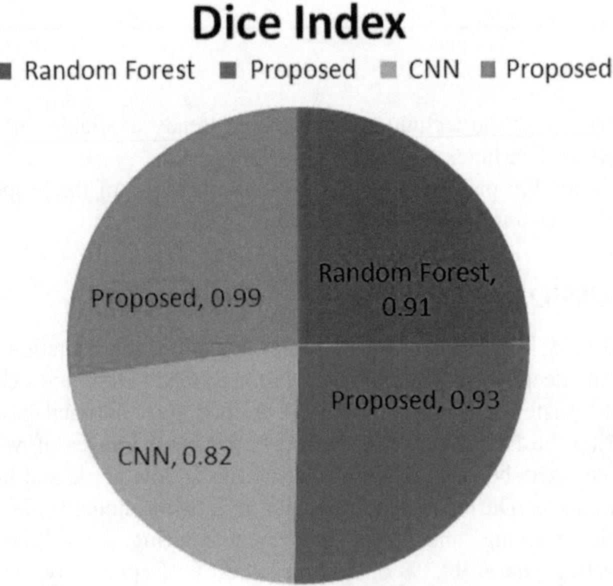

FIGURE 9.6 Performance Results of Proposed System in Terms of Dice Index.

9.6 CONCLUSION AND FUTURE WORK

Automatic brain tumor detection, segmentation and grading are thus implemented by using fully convolutional network and transfer learning techniques using VGG-19 network to the MRI images. The experimental comparison of the proposed system with that of the existing techniques helps to identify that the implemented transfer learning using VGG-19 network outperforms the existing techniques like CNN, Random Forest, etc. The accuracy of the VGG-19 network is improved by deploying future fusion technique, thus, increasing the performance of the system. The scope of the proposed system shall be extended by including the enhancement of the feature map construction by taking into account texture and shape features. Classification accuracy can be improved further by adjusting the network's fully connected and dropout layers. Concatenation of features can be improved to give better outcomes.

REFERENCES

[1]. Wacker J., Ladeira M. and Nascimento J. E. V., "Transfer Learning for Brain Tumor Segmentation", 2019.

[2]. Ishita B., "Brain Tumor Image Segmentation and Classification Using SVM, CLAHE and ARKFCM", Intelligent Decision Support Systems, Applications in Signal Processing, pp. 53–70, 2019, De Gruyter Publication, Germany.

[3]. Havaei M., Davy A., Warde-Farley D., Biard A., Courville A., Bengio Y., Pal C., Jodoin P. M. and Larochelle H., "Brain Tumor Segmentation with Deep Neural Networks", Medical Image Analysis, 35, pp. 18–31, 2017.

[4]. Kamnitsas K., Ferrante E., Parisot S., Ledig C., Nori A. V., Criminisi A., Rueckert D. and Glocker B., "Deep-Medic for Brain Tumor Segmentation", Int. Workshop on Brain Lession – Glioma, Multiple Sclerosis, Stroke and Traumatic Brain Injuries, Springer, pp. 138–149, 2016.

[5]. Ronneberger O., Fischer P. and Brox T., "U-Net Convolutional Networks for Biomedical Image Segmentation", Int. Conf. on Medical Image Computing and Computer Assisted Intervention, Springer, pp. 234–241, 2015.

[6]. Long J., Shelhamer E. and Darrell T., "Fully Convolutional Networks (FCN) for Semantic-Segmentation", 2015.

[7]. Noh H., Hong S. and Han B., "Learning De-Convolution Network for Semantic Segmentation", IEEE Int. Conf. on Computer Vision, pp. 1520–1528, 2016.

[8]. Chen H., Dou Q., Yu L. and Heng P. A., "VoxResNet – Deep Voxelwise Residual Networks for Volumetric Brain Segmentation, pp. 1–9, 2016.

[9]. Kamnitsas K., Ledig C., Newcombe V. F., Simpson J. P., Kane A. D., Menon D. K., Rueckert D. and Glocker B., "Efficient Multi-Scale 3D CNN with Fully Connected CRF for Accurate Brain Lesion Segmentation", Medical Image Analysis, 36, pp. 61–78, 2017.

[10]. Cicek O., Abdulkadir A., Lienkamp S. S., Brox T. and Ronnenberger O., "3D U-Net – Learning Dense Volumetric Segmentation from Sparse Annotation", LNCS, pp. 424–432, 2016.

[11]. Milletari F., Navab N. and Ahmadi S. A., "V-Net: Fully Convolutional Neural Networks (CNN) for Volumetric Medical Image Segmentation", Int. Conf. on 3D Vision, pp. 565–571, 2016.

[12]. Oquab M., Bottou L., Laptev I. and Sivic J., "Learning and Transferring Mid-Level Image Representations Using CNN", IEEE Int. Conf. on Computer Vision and Pattern Recognition, 2014.

[13]. Russakovsky O., Deng J., Su H., Krause J., Satheesh S., Ma S., Huang Z., Karpathy A., Khosla A., Bernstein M., Berg A. C. and Fei-Fei L., "Image-Net Large Scale Visual Recognition Challenge", International Journal of Computer Vision, 115, No. 3, pp. 211–252, 2015.

[14]. Gibson E., Li W., Sudre C. H., Fidon L., Shakir D., Wang G., Eaton-Rosen Z., Gray R., Doel T., Hu Y., Whyntie T., Nachev P., Barratt D. C., Ourselin S., Cardoso M. J. and Vercauteren T., "Nifty-Net: A Deep-Learning Platform for Medical Imaging", Computer Methods and Programs in BioMedicine, 158, pp. 113–122, 2018.

[15]. Madhumathy P., Prasanth G., Rashmi B. R. and Sri Lekha, "Eye Movement Detection for Paralyzed Patient Using Pressure Sensor", International Journal of Scientific & Engineering Research, 7, No. 11, pp. 395–400, 2016.

[16]. Sudiatmika I. B. K., Rahman F. and Trisno S., "Image Forgery Detection using Error-Level Analysis and Learning", TELKOMNIKA, 17, No. 2, pp. 53–659, 2019.

[17]. Li Y., Qi H., Dai J., Ji X. and Wei Y., "Fully Convolutional Instance Aware Semantic-Segmentation", IEEE Conf. on Computer Vision and Pattern Recognition, pp. 4438–4446, 2017.

[18]. Pereira S., Pinto A., Alves V. and Silva C. A., "Brain Tumor Segmentation Using CNN in MRI Images", IEEE Transactions on Medical Imaging, 35, No. 5, pp. 1240–1251, 2016.

[19]. Tran P. V., "A Fully CNN for Cardiac Segmentation in Short Axis MRI", arxiv – 1604.00494, 2016.

[20]. Alaskar S., Abdullah M. A. M. and Jebur B. A., "Automatic Brain Tumor Segmentation Using Fully CNN and Transfer Learning", ICECCPEC19, pp. 188–192, 2019.

[21]. Ahuja S., Panigrahi B. K. and Gandhi T., "Transfer Learning Based Brain Tumor Detection and Segmentation Using Super-Pixel Technique", Int. Conf. on Contemporary Computing and Applications, pp. 244–249, 2020.

[22]. Naser M. A. and Jamal Deen M., "Brain Tumor Segmentation and Grading of Low Grade Glioma (LGG) Using Deep Learning in MRI Images", Computers in Biology and Medicine, 121, pp. 1–8, 2020.

[23]. Pravitasari A. A., Iriawan N., Almuhayar M., Azmi T., Irhamah, Fithriasari K., Purnami S. W. and Ferriastuti W., "UNet-VGG16 with Transfer Learning for MRI Based Brain Tumor Segmentation, TELKOMNIKA, 18, No. 3, pp. 1310–1318, 2020.

[24]. Talo M., Baloglu U. B., Yildirim O. and Acharya U. R., "CNN for Multi-Class Brain Disease Detection Using MRI Image", Computerized Medical Imaging and Graphics, 78, pp. 1–12, 2019.

[25]. Sajid S., Hussain S. and Sarwar A., "Brain Tumor Detection and Segmentation in MR Images Using Deep Learning", Arabian Journal for Science and Engineering, 11, pp. 9249–9261, 2019.

10 Deep Learning-Computer Aided Melanoma Detection Using Transfer Learning

Mohan Kumar S.[1], *T. Kumanan*[2], *Dr. T. R. Ganesh Babu*[3], *and S. Poovizhi*[4]
[1]Professor, Nagarjuna College of Engineering and Technology
[2]Principal, Faculty of Engineering and Technology, Meenakshi Academy of Higher Education and Research
[3]Professor, Department of ECE, Muthayammal Engineering College
[4]Research Scholar, Anna University

CONTENTS

10.1 INTRODUCTION

A research conducted by the World Health Organization (WHO) shows that Melanoma is a type of skin cancer which is common among adults between the ages

DOI: 10.1201/9781003194415-10

of 25 and 29 years and accounts for 75% of fatalities from cancer related causes. But research proves that if detected in time, this proactiveness may help in increasing the chances of survival of the patient [1]. Given the main lead of skin cancer is the fact that it generally occurs in the skin cells which are the external cells and is easily visible to the human eye, therefore, self-examination which is conducted by a particular person or even by a trained physician can help in early detection. In addition, the fact that melanoma spreads throughout the body very fast by means of lymphatic nodes makes it very dangerous and emphasizes the need for faster detection which, otherwise, may prove fatal. Since our skin acts as a protective covering for our body and shields us from various deleterious effects like sun's heat, sunlight, any injury, and skin infection, it's in fact excessive exposure of our skin to the UV rays from the sun that leads to skin cancer. The fair-complexioned population has been more at risk from melanoma [2]. Diagnosis at an early stage improves the chances of proper treatment and cure which otherwise may be fatal. If we take into account that each year close to 55,000 people are diagnosed with this type of cancer, any significant change to an already present mole in terms of physiological appearances is a warning sign that must not be ignored. Dermatoscopy is a novel non-invasive technique that enables the magnification of the skin cells and enables a visual inspection of the underlying skin which provides a clear image of spots and moles on the skin [3]. Recent development in new technology like machine learning and image processing made a remarkable revolution in the field of medical sciences. Specifically, Computer aided Diagnosis (CAD) of diseases is a term coined for computer-based diagnosis of diseases [4,5].

The skin protects our body and shields us from the sun's heat, sunlight, any injury, and skin infection. However, excessive exposure of our skin to the UV rays from the sun leads to skin cancer. The fair-complexioned population has been more at risk of developing melanoma. Given diagnosis at an early stage improves the chances of proper treatment and cure which, otherwise, may be fatal, it's a matter of no little concern that each year close to 55,000 people are diagnosed with this type of cancer. The skin is the largest portion of the human body, and it protects the internal parts of the body from harmful UV radiation from the sun. Further, the human skin is made up of mostly three layers which are the dermis, the epidermis and the hypodermis [6]. Melanin as the skin coloring pigment is a main element of the skin and excessive production of this pigment may lead to skin cancer. Though occurrence of brown spots and normal growths on the skin are usually harmless and may be non-carcinogenic, still excessive exposure to UV rays may lead to uncontrolled multiplication of the skin cells which is a characteristic of melanoma. Those who have over 100 moles all over their body are at a higher risk [7,8]. The moles have to be observed for the ABCD signs or, the asymmetry, border, Color and Diameter.

- Asymmetry: The non-carcinogenic moles are usually symmetrical compared to malignant moles that are asymmetrical [9].
- Borders: The non-carcinogenic moles generally have very smooth and even borders compared to malignant moles which are non-uniform [10].
- Color: The non-carcinogenic moles generally possess a single color compared

to malignant moles that have a variety of colors and this is a sign that is of great significance [11].

- Diameter: The non-carcinogenic moles generally are smaller and may be less than 6 mm in diameter compared to larger moles that might be malignant [12]. Figure 10.1 collates and compares non-carcinogenic versus malignant moles by their signs.

Any significant change to an already present mole in terms of physiological appearances is a warning sign that must not be ignored, and for the purpose of detecting early warning signs, dermatoscopy is a novel non-invasive technique that enables the magnification of the skin cells and enables a visual inspection of the underlying skin which provides a clear image of spots and moles on the skin.

With the advancements in machine learning and image processing, computer-based diagnosis of skin cancer has set in motion a revolution in the analysis of identifying cancer [13]. The digital dermascopes are used to capture images of the skin, and these images may be segmented in order to isolate the pigmented area; and is followed by extraction of significant features which are then used by trained classifiers to identify melanoma [14,15]. Even when an accomplished dermatologist does the dermoscopy for diagnosis, still the accuracy of melanoma recognition is about 75–84%. But, computer aided diagnosis (CAD) ensures better results as far as accuracy and speed of diagnosis is concerned. Although compute abilities of machine facilities cannot be termed as brilliant compared to those of the human's, anyhow, computer makes the diagnosis efficient by identifying varying color in the skin, irregularity and texture features which may not be visualized easily by human eyes [16].

The digital dermoscopes are used to capture images of the skin, and these images may be segmented in order to isolate the pigmented area, followed by extraction of significant features which are then used by trained classifiers to identify melanoma [17,18]. An accomplished dermatogist can estimate melanoma about 75–84% of the times he scans images for the purpose. Computer diagnosis provides better accuracy and speedy process.

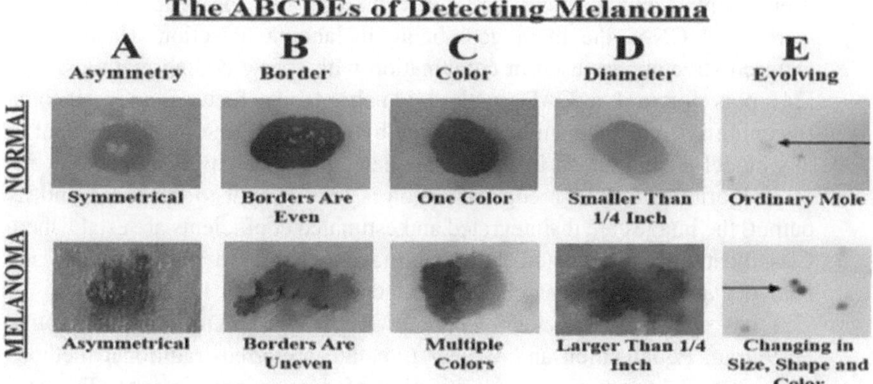

FIGURE 10.1 ABCD features for skin detection.

In recent years, image processing techniques have been suggested by many researchers for diagnosing melanoma. Image processing plays a vital role to enhance image quality by reproducing digital images with good brightness/contrast, where detail is a strong requirement in the medical field. Digital image acquisition is an important requirement to be performed on digital images [19]. In this process, every pixel is assigned with a label, which will share same visual behaviors.

10.2 RELATED RESEARCH WORK

An extensive literature survey of the already present methods was conducted and it was noticed that different authors have suggested various ways and means to improvise melanoma detection.

- [20] has suggested a system employing hybrid spatial feature by means of the PNN classifier and uses representation and radial basis type network classifier to distinguish between normal and abnormal skin lesions and further classify the same.
- [21] proposed a CAD system for melanoma detection. Markov and Laplace Filter is further used to eradicate non-essential elements and, lastly, the lesion edge is detected. The RGB image is converted to YUV color space and, then, Channel-U is chosen for processing. The conversion of the image to binary form is done by Otsu's thresholding method.
- [22] has suggested CAD system for detecting skin cancer. A set of dermoscopic images serves as input to the system. In this particular system, extract gray-level co-occurrence matrix features and, for classification, multilayer perceptron classifier is used.
- This classifier utilizes two different techniques in the process of training and testing:
    - automatic multilayer perceptron classifier
    - traditional multilayer perceptron classifier.

- [23] suggests a novel way for automated detection of skin cancer combining deep learning and machine learning techniques like the deep residual networks and CNN, the main goal being melanoma detection. Hand coded feature extractors are used in combination with sparse coding methods.
- [24] puts forward a CAD method which mainly focuses on melanoma screening systems that can be used even by untrained personnel. The input to this model is skin lesion images. Edge detection is achieved by contour tracing algorithm. After the edge detection is fed as input to the DWT and its output, the images are disintegrated and estimated coefficients are established. Classification of normal and abnormal images is achieved by the Probabilistic Network and the K-means clustering algorithm.
- [25] has suggested a system where in the preprocessing stage, Adaptive Histogram Equalization and Weiner filter are preferred traditional methods planned for separation and classification of dermoscopic images. The segmentation is performed using active contour segmentation. Texture features

entropy, correlation, homogeneity and energy features are extracted. Thereafter, separate the images into first, second and higher order features, and SVM classifier is suggested for feature classification.

- [26] Mansoura has put forward a CAD system for identification of dermoscopic images using machine learning techniques. At first, the image segmentation is performed. Using fuzzy C means, the fuzzy entropy and morphology based optical mask selection is done; based on optimal mask, the adaptive contour method segments skin lesions; and thereafter, there is refinement of segmentation using morphology operations.

Generally, skin lesion is considered as doubtful in diagnosing melanoma. The images are subject to preprocessing to eject hair and noise etc. which is helpful in obtaining an image of very good quality [27,28]. Given the Modified Otsu thresholding algorithm is ideal for image segmentation, the various algorithms, for example, border irregularity, spread area, axis length, etc. are used for feature classification stage; and the extracted features are processed with the intention of classifying the image as mole, benign, suspicious or highly suspicious skin lesions [29]. The following steps are carried out in a diagnosis of melanoma skin cancer:

1. Acquisition of image lesion.
2. Preprocessing of image to reduce noise and various irregularities.
3. Segmentation of the lesion area carried from the outer area.
4. Extraction features from input image.
5. Classification based on the extracted features.

Feature extraction characterized by identifying specific features or characters that are primarily responsible for melanoma detection. Frequently, all skin lesions must not have all the ABCD feature (combination of features like e.g. A + B, A + C, B + C, A + B + C, etc.) [30,31]. Figure 10.2 block diagram is used to pictorially depict the general outline of methodology for melanoma detection.

In Figure 10.3a and 10.3b, we consider the images of skin lesions before and after preprocessing is performed. In Figure 10.3c and 10.3d, the histograms of original images and the preprocessed images are constructed and compared accordingly.

10.2.1 Benign Sample Image 1

10.2.2 Benign Sample Image 2

In Figure 10.4a and 10.4b, we consider the images of skin lesions before and after preprocessing is performed. In Figure 10.4c and 10.4d, the histograms of original images and the preprocessed images are constructed and compared accordingly.

The above histograms clearly show the difference in the pixel density of colors between the original image and the preprocessed image.

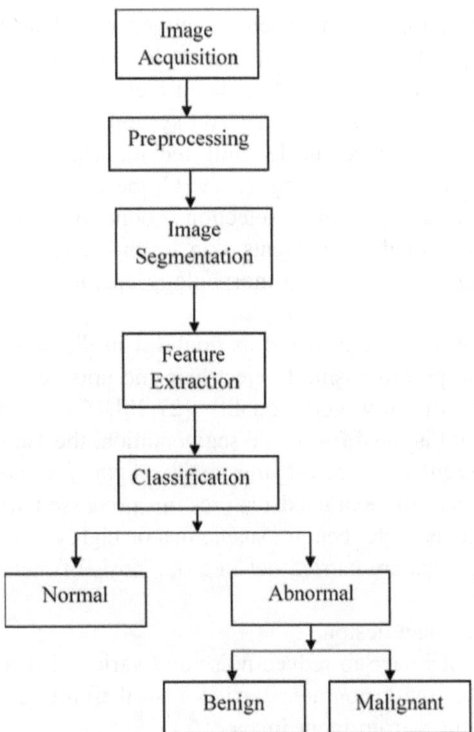

FIGURE 10.2 Block diagram.

10.2.3 Melanoma Sample Image 1

In Figure 10.5a and 10.5b, we show the images of skin lesions before and after preprocessing is done. In Figure 10.5c and 10.5d, we show the histograms of the image before and after preprocessing is done.

10.2.4 Melanoma Sample Image 2

In Figure 10.6a and 10.6b, the image of skin lesions before and preprocessing is done is depicted. In Figure 10.6c and 10.6d, the histograms of the original image before and after preprocessing is done are depicted.

The histograms clearly show the difference in the pixel density of colors between the original image and the preprocessed image.

10.3 TRANSFER LEARNING CAD SCC MODEL

ISIC dataset has been taken for training this model which has 1.8k benign and 1.4k malignant images. Model was trained using Transfer Learning technique. InceptionV3 was used as base network. InceptionV3 has 311 layers. The model was trained by retraining 100th layers onwards [32,33]. GlobalAveragePooling followed

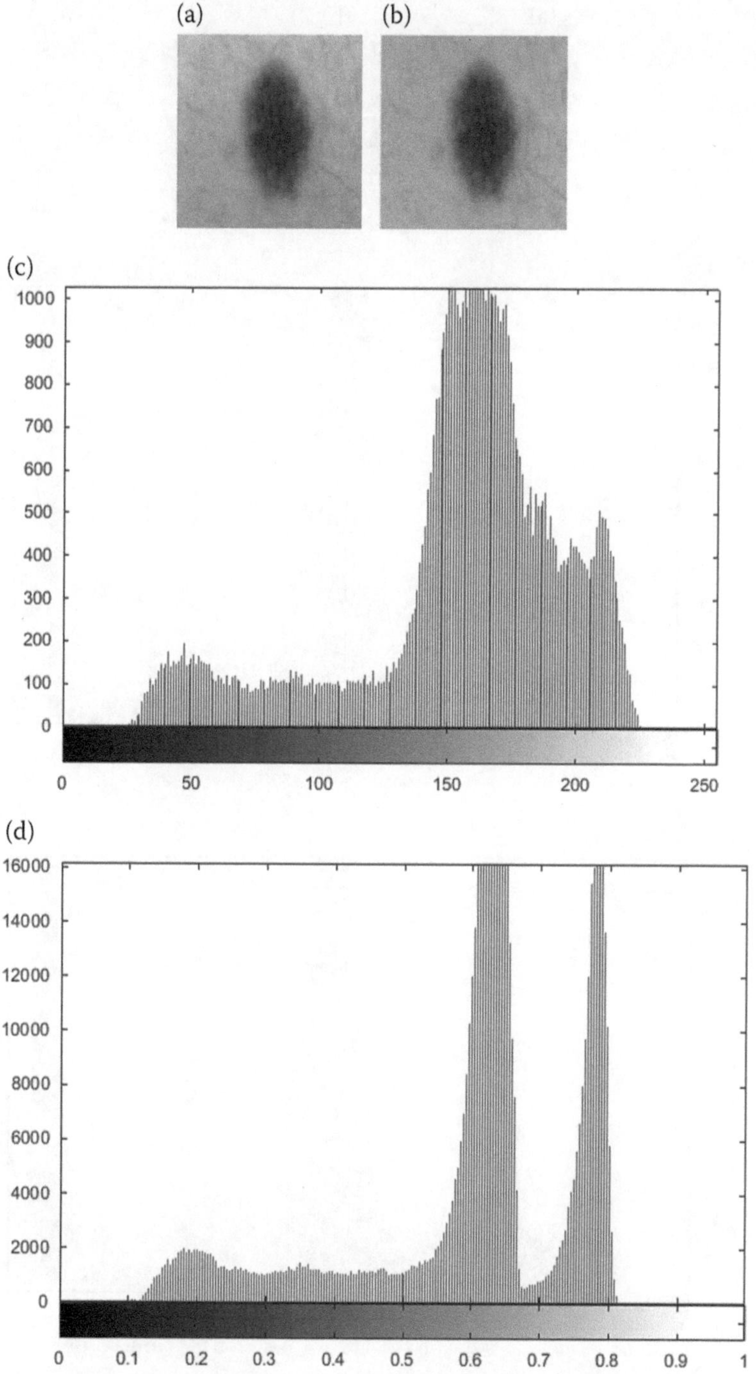

FIGURE 10.3 (a) Original Image, (b) image after preprocessing, (c) histogram of original image, (d) histogram of preprocessed image.

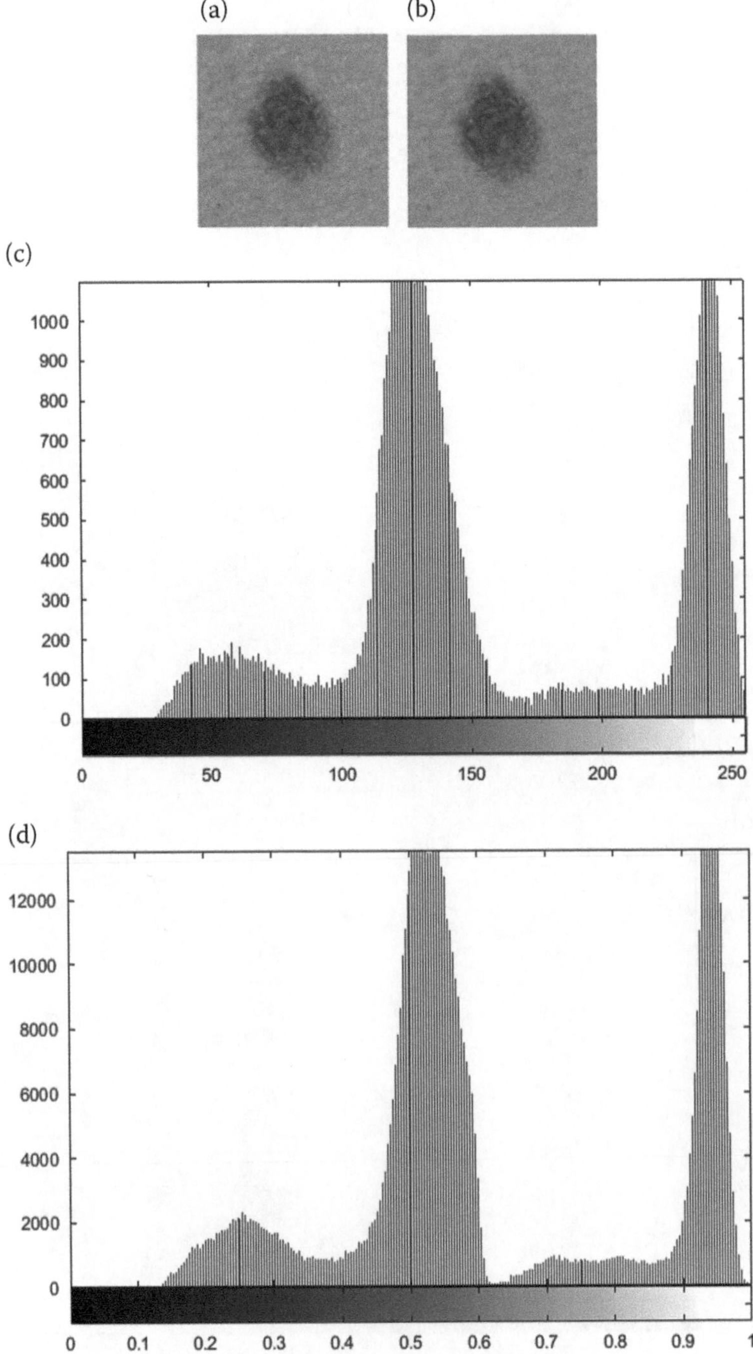

FIGURE 10.4 (a) Original Image, (b) image after preprocessing, (c) histogram of original image, (d) histogram of preprocessed image.

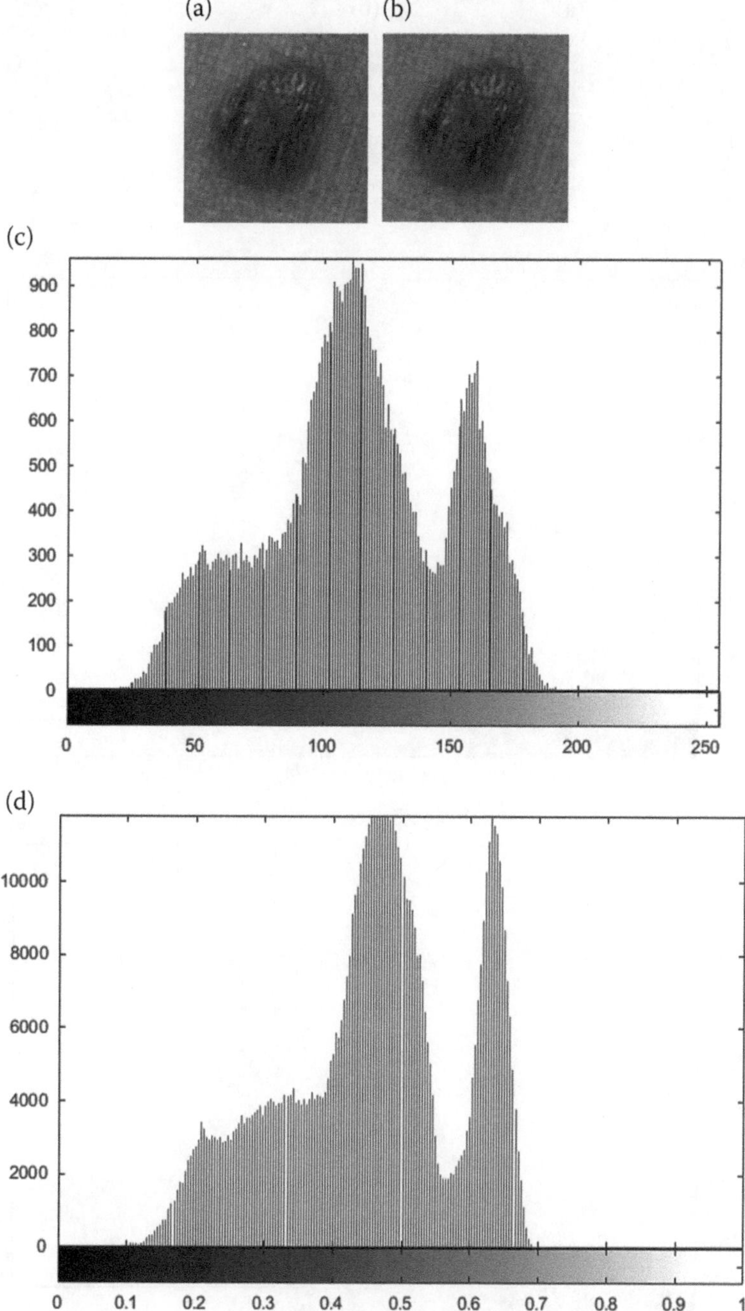

FIGURE 10.5 (a) Original Image, (b) image after preprocessing, (c) histogram of original image, (d) histogram of preprocessed image.

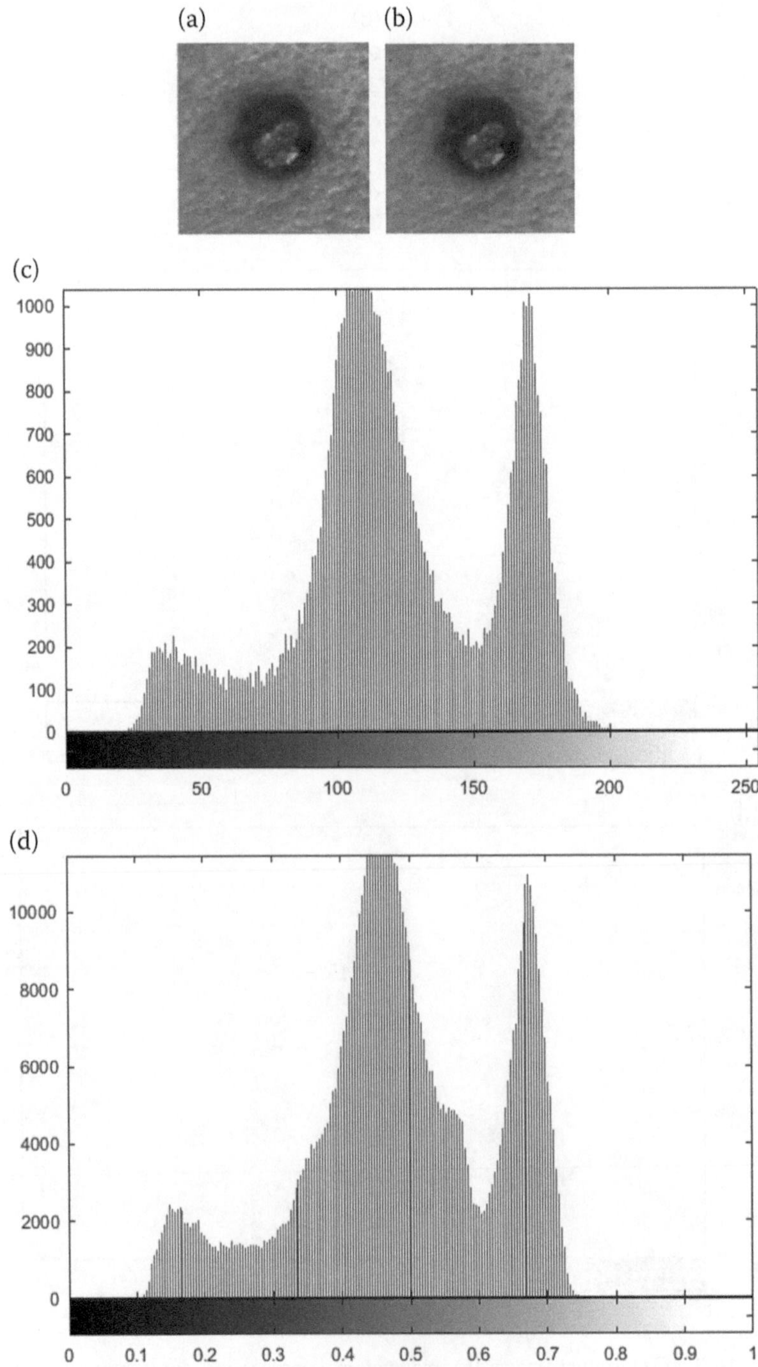

FIGURE 10.6 (a) Original Image, (b) image after preprocessing, (c) histogram of original image, (d) histogram of preprocessed image.

by output layer with SoftMax is used. Input images are resized to 299 × 299 × 3 as depicted in Figure 10.7.

10.3.1 Model Summary

Model was trained with 21m trainable parameters.

Total parameters: 21,810,980
Trainable parameters: 21,776,548
Non-trainable parameters: 34,432

10.3.2 Sample Images

Figure 10.8 depicts the image datasets which are being trained and tested by the melanoma detection system using Transfer Learning.

10.4 ACCURACY RESULTS ACHIEVED THROUGH THE PROPOSED PROCESSING

Figure 10.9 depicts the train accuracy level and validation accuracy level achieved by the melanoma detection System using Transfer Learning.

10.4.1 Loss Results Achieved Through the Proposed Processing

Figure 10.10 depicts the train loss level and validation loss level of the melanoma detection system using Transfer Learning.

10.4.2 Confusion Matrix

The heatmap of the Confusion matrix of Validation Data is given in Figure 10.11. Out of 360 benign samples, 327 were predicted correctly; and out of 300 malignant samples, 272 were predicted as malignant.

10.4.3 Classification Report

Table 10.1 tabulates the classification results for abnormal images classified into benign, malignant, accuracy of classification, macro average and weighted average based on precision, recall, F1-score and support.

10.5 CONCLUSION

From the proposed model, a recall score of 91% is achieved. As the medical systems are high-recall systems, which means a person with malignant cancer should not be predicted as benign case as this leads to the person's plausibly not going for further treatment. With image augmentation and the taking of more samples of the images, better recall score shall be achieved in future.

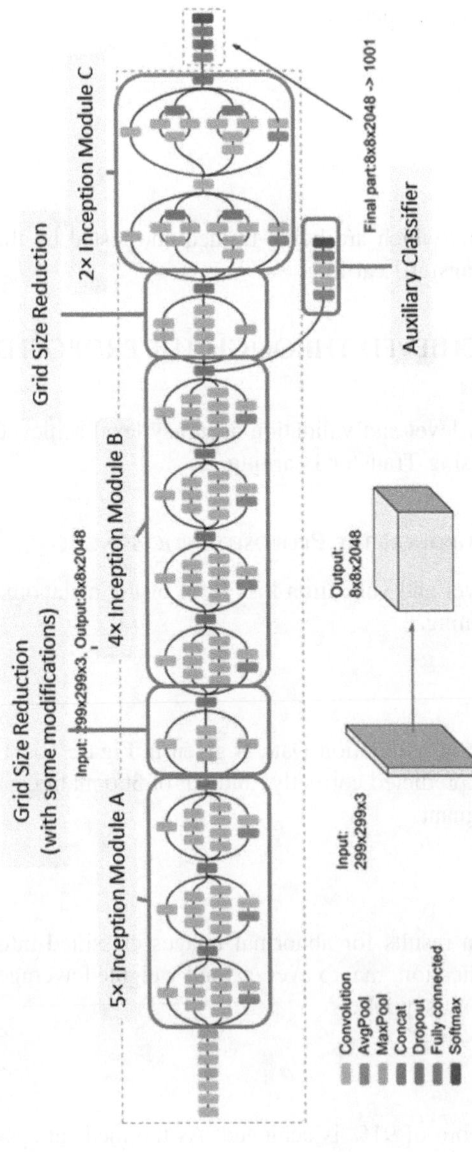

FIGURE 10.7 Model of InceptionV3.

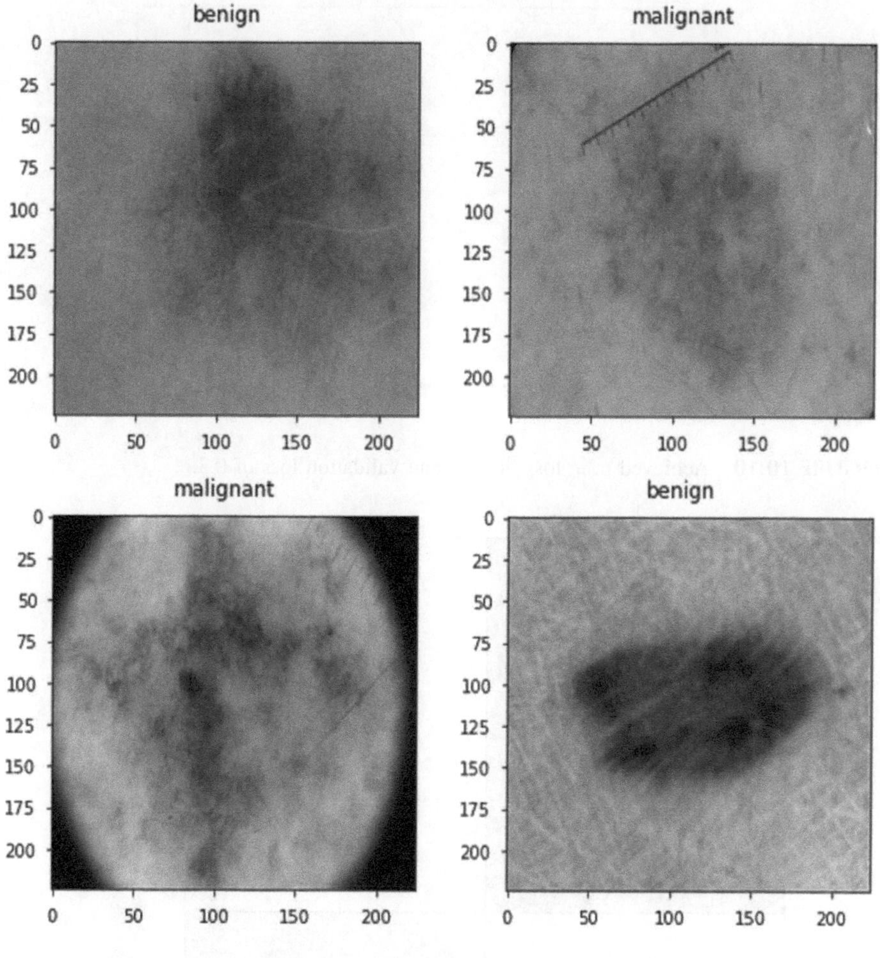

benign

malignant

malignant

benign

FIGURE 10.8 Melanoma visuals.

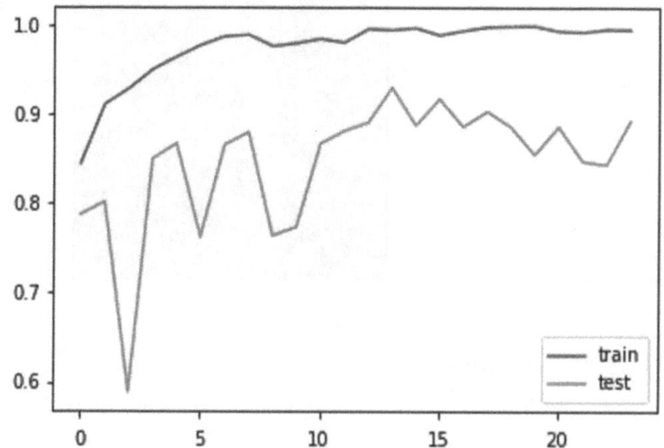

FIGURE 10.9 Achieved train accuracy of 0.99 and validation accuracy of 0.92.

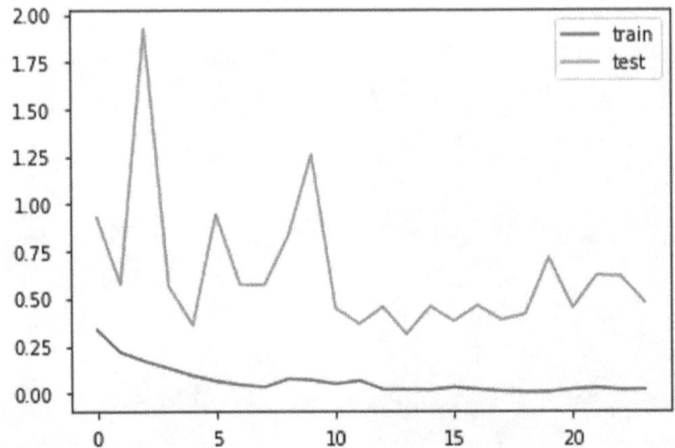

FIGURE 10.10 Achieved train loss of 0.01 and validation loss of 0.3.

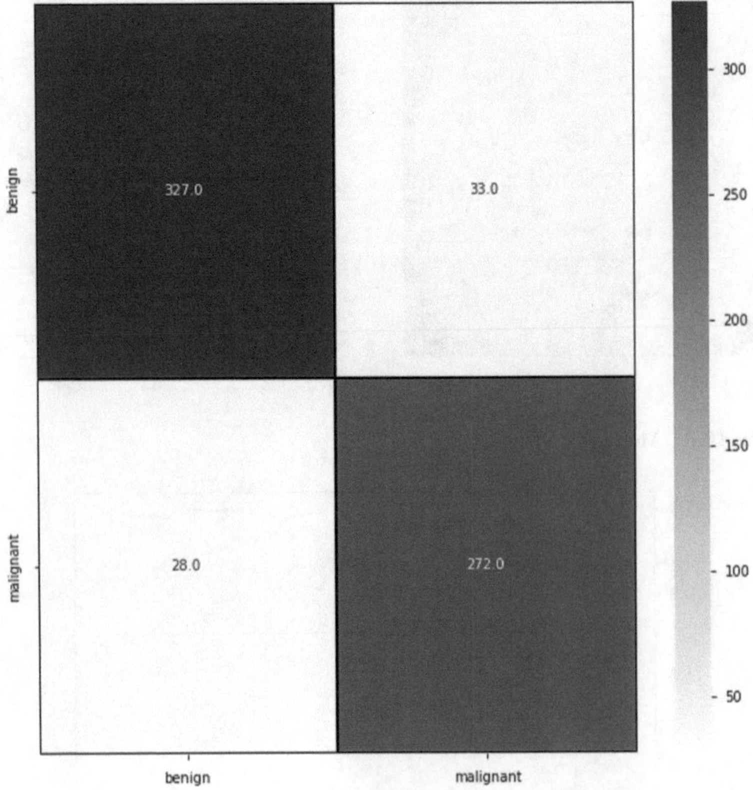

FIGURE 10.11 Confusion matrix.

TABLE 10.1
Classification report

	Precision	Recall	F1-score	Support
Benign	0.92	0.91	0.91	360
Malignant	0.89	0.91	0.90	300
Accuracy			0.91	660
Macro average	0.91	0.91	0.91	660
Weighted average	0.91	0.91	0.91	660

REFERENCES

1. Wu, J., "Efficient Hik SVM Learning for Image Classification", *IEEE Transactions on Image Processing*, Vol. 21, No. 10, October 2012.
2. Wang, F. and Wang, F., "Rapidly Void Detection in TSVS with 2-D X-Ray Imaging and Artificial Neural Networks", *IEEE Transactions on Semiconductor Manufacturing*, Vol. 27, No. 2, May 2014.
3. Bianchini, M. and Scarselli, F., "On the Complexity of Neural Network Classifiers: A Comparison Between Shallow and Deep Architectures", *IEEE Transactions on Neural Networks and Learning Systems*, Vol. 25, No. 8, August 2014.
4. Nasir, M., Attique Khan, M., Sharif, M., Lali, I.U., Saba, T., and Iqbal, T., "An Improved Strategy for Skin Lesion Detection and Classification Using Uniform Segmentation and Feature Celection-Based Approach", *Microscopy Research and Technique*, Vol. 81, No. 6, 528–543, 2018.
5. Banerjee, I., "Brain Tumor Image Segmentation and Classification using SVM, CLAHE and ARKFCM", *Intelligent Decision Support Systems, Applications in Signal Processing*, pp. 53–70, De Gruyter, Germany, October 2019.
6. Lu, L., Di, L., Sr and Ye, Y., "A Decision-Tree Classifier for Extracting Transparent Plastic-Mulched Landcover from Landsat-5 TM Images", *IEEE Journal of Selected Topics in Applied Earth Observations and Remote Sensing*, Vol. 7, No. 11, November 2014.
7. Divya, D. and Ganesh Babu, T.R., "Fitness Adaptive Deer Hunting-Based Region Growing and Recurrent Neural Network for Melanoma Skin Cancer Detection", *International Journal of Imaging Systems and Technology*, Vol. 30, No. 3, 731–752, 2020, ISSN 0899-9457, E-ISSN 1098-1098, DOI: 10.1002/ima.22414.
8. Li, C.-H., Kuo, B.-C., Lin, C.-T. and Huang C.-S., "A Spatial–Contextual Support Vector Machine for Remotely Sensed Image Classification", *IEEE Transactions on Geoscience and Remote Sensing*, Vol. 50, No. 3, March 2012.
9. Ananthajothi, K. and Subramaniam, M., "Multi-level Incremental Influence Measure-Based Classification of Medical Data for Improved Classification", *Cluster Computing*, Vol. 13, No. 5, 783–789, 2018. 10.1007/s10586-018-2498-z.
10. Silva, C.S. and Marcal, A.R., "Colour-Based Dermoscopy Classification of Cutaneous Lesions: An Alternative Approach", *Computer Methods in Biomechanics and Biomedical Engineering: Imaging & Visualization*, Vol. 1, No. 4, 211–224, 2013.
11. Mendonça, T., Ferreira, P.M., Marques, J.S., Marcal, A.R. and Rozeira, J., "PH 2-A Dermoscopy Image Database for Research and Benchmarking", in 2013 35th Annual

International Conference of the IEEE Engineering in Medicine and Biology Society (EMBC) (pp. 5437–5440), July 2013. IEEE.
12. Yap, J., Yolland, W. and Tschandl, P., "Multimodal Skin Lesion Classification Using Deep Learning", *Experimental Dermatology*, Vol. 27, No. 11, 1261–1267, 2018.
13. Jaworek-Korjakowska, J. and Tadeusiewicz, R., "Determination of Border Irregularity in Dermoscopy Color Images of Pigmented Skin Lesions", in Conference Proceedings: Annual International Conference of the IEEE Engineering in Medicine and Biology Society. IEEE Engineering in Medicine and Biology Society. Annual Conference (Vol. 2015, pp. 2665–2668), August 2015.
14. Poovizhi, S. and Ganesh Babu, T.R., "An Efficient Skin Cancer Diagnostic System Using Bendlet Transform and Support Vector Machine", *Annals of Brazilian Academy of Sciences*, Vol. 92, No. 1, 1–12, 2020.
15. Jain, Y.K., "Comparison Between Different Classification Methods with Application to Skin Cancer", *International Journal of Computer Applications*, Vol. 53, No. 11, September 2012.
16. Dash, S., Senapati, M.R. and Jena, U.R., "K-NN Based Automated Reasoning Using Bilateral Filter-Based Texture Descriptor for Computing Texture Classification", *Egyptian Informatics Journal*, Vol. 19, No. 2, 133–144, 2018.
17. Ruela, M., Barata, C., Marques, J.S. and Rozeira, J., "A System for the Detection of Melanomas in Dermoscopy Images Using Shape and Symmetry Features", *Computer Methods in Biomechanics and Biomedical Engineering: Imaging & Visualization*, Vol. 5, No. 2, 127–137, 2017.
18. Umbaugh, S.E., Moss, R.H., Stoecker, W.V. and Hance, G.A., "Automatic Colour Segmentation Algorithms with Application to Skin Tumour Feature Identification", *IEEE Engineering in Medicine and Biology*, Vol. 12, 75–82, 1993.
19. Barhoumi, W. and Baâzaoui, A., "Pigment Network Detection in Dermatoscopic Images for Melanoma Diagnosis", *IRBM*, Vol. 35, No. 3, 128–138, 2014.
20. El Abbadi, N.K. and Faisal, Z., "Detection and Analysis of Skin Cancer from Skin Lesions", *International Journal of Applied Engineering Research*, Vol. 12, No. 19, 9046–9905, 2017.
21. Codella, N., Nguyen, Q.-B., Pankanti, S., Gutman, D., Helba, B., Halpern, A. and Smith, J.R., "Deep Learning Ensembles for Melanoma Recognition in Dermoscopy Images", *IBM Journal of Research and Development*, Vol. 61, 1–15, 2017. 10.1147/JRD.2017.2708299.
22. Sheha, M.A., "Automatic Detection of Melanoma Skin Cancer Using Texture Analysis", *International Journal of Computer Applications*, Vol. 42, No. 20, 22–26, 2012.
23. Mabrouk, M.S., Sheha, M.A. and Sharawy, A., "Automatic Detection of Melanoma Skin Cancer Using Texture Analysis," *International Journal of Computer Applications*, Vol. 42, No. 20, 22–26, 2012.
24. Argenziano, G., Soyer, H.P., Chimenti, S., Talamini, R., Corona, R., Sera, F., et al., "Dermoscopy of Pigmented Skin Lesions: Results of a Consensus Meeting Via the Internet", *Journal of the American Academy of Dermatology*, Vol. 48, 679–693, 2003.
25. Saida, T., Miyazaki, A., Oguchi, S., Ishihara, Y., Yamazaki, Y. and Murase, S., "Significance of Dermoscopic Patterns in Detecting Malignant Melanoma on Acral Volar Skin", *Archives of Dermatology*, Vol. 140, 1233–1238, 2004.
26. Jayapal, P., Manikandan, R., Ramanan, M., Shiyam Sundar, R.S. and Udhaya Suriya, T.S., "Skin Lesion Classification Using Hybrid Spatial Features and Radial Basis Network", *International Journal of Innovative Research in Science, Engineering and Technology*, Vol. 3, No. 3, 10014–10021, 2014.
27. Fatima, R., Khan, M.Z.A., Govardhan, A. and Dhruve, K.D., "Computer Aided

Multi-parameter Extraction System to Aid Early Detection of Skin Cancer Melanoma", *International Journal of Computer Science and Network Security*, Vol. 12, No. 10, 74–86, October 2012.

28. Jeya Ramya, V., Navarajan, J., Prathipa, R. and Ashok Kumar, L., "Detection of Melanoma Skin Cancer Using Digital Camera Images", *ARPN Journal of Engineering and Applied Sciences*, Vol. 10, No. 7, 3082–3085, April 2015.
29. Argenziano, G., Fabbrocini, G., Carli, P., De Giorgi, V., Sammarco, E. and Delfino, M., "Epiluminescence Microscopy for the Diagnosis of Doubtful Melanocytic Skin Lesions. Comparison of the ABCD Rule of Dermatoscopy and a New 7-Point Checklist Based on Pattern Analysis", *Archives of Dermatology*, Vol. 134, No. 12, 1563–1570, 1998.
30. Korotkov, K. and Garcia, R., "Computerized Analysis of Pigmented Skin Lesions: A Review", *Artificial Intelligence in Medicine*, Vol. 56, 69–90, 2012.
31. Altamura, D., Altobelli, E., Micantonio, T., Piccolo, D., Fargnoli, M.C. and Peris, K., "Dermoscopic Patterns of Acral Melanocytic Nevi and Melanomas in a White Population in Central Italy", *Archives of Dermatology*, Vol. 142, 1123–1128, 2006.
32. Yu, L., Chen, H., Dou, Q., Qin, J. and Heng, P.A., "Automated Melanoma Recognition in Dermoscopy Images Via Very Deep Residual Networks", *IEEE Transactions on Medical Imaging*, Vol. 36, 994–1004, 2017.
33. AlMansour, E., Jaffar, M.A. and AlMansour, S., "Fuzzy Contour Based Automatic Segmentation of Skin Lesions in Dermoscopic Images", *International Journal of Computer Science and Network Security*, Vol. 17, No. 1, 177–186, January 2017.

11 Development of an Agent-Based Interactive Tutoring System for Online Teaching in School Using Classter

Dr. Tribhuwan Kumar[1]*, Digvijay Pandey*[2]*, and Dr. R Umamaheshwari*[3]

[1]Assistant Professor of English, College of Science and Humanities at Sulail, Prince Sattam Bin Abdulaziz University

[2]Department of Technical Education, IET, Dr. A.P.J. Abdul Kalam Technical University

[3]Assistant Professor, Department of Electronics and Instrumentation Engineering, SRM Valliammai Engineering College

CONTENTS

DOI: 10.1201/9781003194415-11

11.1 INTRODUCTION

'E-learning is highly useful to the students and teachers. It can use various e-learning technologies and resources to bring innovation in teaching and learning process' [1,2]. In education system and situation research, the standard interactive tutoring systems (SILS) and personalization are well thought-out with the two major significant factors. First, Artificial interactive tutoring systems serve up at the same time as an effective instrument to get better predicament solve skill by simulating a human tutor's action [3]. By simulating, a human tutor's action implements conversation adaptive and personalized teaching skills. Therefore, in this goal, an interactive tutoring system is proposed. Second, it reimbursement as of RNN algorithm in organization ambiguity based on the likelihood theory designed for the procedure of student management to give support to learn online teaching programs. In addition, the standard-based intelligent tutoring system reimbursement as of a multi-agent scheme employs a routine teach-to-write exchange model. This is dovetailed with a beginner's online teaching programs as well as with the Classter tool to develop the student's problem-solving skills. Last, the presentation of Standard Intelligent Tutoring System is investigated from beginning to end as an experimental study. It is revealed that SILS is a tool to improve students' tutoring attention, approach and knowledge receipt, but it may as well help student to attain supplementary capability in stipulation of problem-solving behaviors [4].

Classter is an influential, modern web-based school management system and it facilitates the overall computerization of all operations of your educational institute. It combines management, tutoring, teaching and assessment, bringing each department and every stakeholder to work on a single platform. Classter consent is for uncomplicated administration of candidates, admission and registration, classes, courses and English subjects, curriculums and educational plans. Given Classter provides the web portals for any user type, it is of assistance for the organization to collect, supervise and analyze student information in sequence, access their data and monitor their progress, behavior and achievements [5,6].

The clatter site has strength of character to keep in custody the information's in sequence concerning the students [7]. To start off, the student's registration at the site is essential for their assessment acquiescence through the system. Initial registers make available the delicate details keen on the system. Then the website makes use of the data and get to the front end of the information. At the time of registration, the users are required to provide their full name, date of birth, university, username and password kept on the system. The website resolves the resolute values of beginner as in level 1 student. This stage is, on the other hand, not stable as the user level will probably vary depending on their development [8]. There are three sets of student levels that are documented in this method: level 1 or learner stage, level 2 or in-between stage, and level 3 or progressive level. The student may possibly be present in any level as well as in their level based on their progress produced by the system. In this paper, 665 students of matriculation school from the Bihar state are taken into account. These students are selected based on their being tutored their English subjects through the web-based online teaching portal Classter.

11.2 LITERATURE REVIEW

Intelligent Tutoring System is categorized into the following four basic models taking from the mechanism based on a wide-ranging agreement in the middle of researcher [9]. The models are: Domain representation, Student representation, Tutoring representation, and User interface representation. For modeling domain values, the domain model is used, and is also called the cognitive model, the mode of specialized knowledge. The domain model is designed on the ACT-R theory of education and in this way it applies to learning. It attempts to have a clear understanding of the probable steps needed to solve a problem [10]. Especially, domain representation consists of the concept, regulations, and investigative strategy of the domain which the classes are tutoring. They will provide the knowledge of the expert in their tutoring English subjects, a benchmark for estimating the student's presentation or for becoming aware of error [11].

A standard interactive tutoring system (SILS) is introduced taking into consideration the student's study and superimposes resting on the domain model. It is conceptually well-crafted, while being an essential component of a SILS that caters to students' cognitive and emotional states. And their development of tutoring process is made advanced in this domain. At the same time, the students must work gradually from beginning to end on their problem-solving procedure [12]. An SILS fits into place in an expansion call model tracing. At any time, the instance the students' models deviate from their domain model, this arrangement identify, or streams, avoid the error occurrences. Alternatively, the constraint-based online teacher develops the student model which represents and superimposes on top of the constraint set [13]. The student outcomes or solutions in opposition to the constraint set are also recognized, satisfied and dishonored by the constraint-based tutor. If any constraint is violated, the attained results of the student's results are wrong [14–16]. In addition, the SILS make available the comment based on errors on those constraints. Constraint-based tutor provide feedbacks over errors in addition to some optimistic feedback too.

In view of SILS the fact is the IL technology skills are very essential in education process. It increases the expectations for educational institutions, which forces students to make the choice of a specific school a priority. The Learning Resource Management Programs increase substantially the SILS competence [17]. SILS competency combined with the real-time environment has enhanced the classroom teaching. The current scenario shows that the world is becoming digitalized. With this development, teaching and learning processes also improve day by day based on the digital techniques. Also, the implementation of electronic teaching methods improve the students' learning abilities [18]. There are many potential benefits in using SILS, particularly, in distance education such as benefits of time saving, book free study, easy understanding and easy study material access.

Artificial Neural Network (ANN) is the major branch of Artificial Intelligence network system to solve the numerous real-time problems in life. ANN has solved important challenges like real estate, finance, healthcare, education, medicine, pattern recognition, and many more fields [19]. Because of the multilayer perception of ANN, it is able to represent the non-linear function map among input and output. ANN network has been equipped with the powerful yet computationally efficient techniques. It is important to determine the correlation between the input and output of the model in order to properly utilize ANN [20]. ANN studies have been conducted and forecasting of the success of various departments has been conclusively demonstrated.

Because ANN is used to measure student achievement assessments and to track them in teaching and learning programs, it is utilized to assist in teaching and learning [2]. RNN is effective for forecasting several different types of school admission results across various timeframes. The findings show that ANN can increase the effectiveness of a school admission system with an accuracy of seventy percentages [21]. The use of a multilayer perceptional recurrent neural network (RNN) for back propagation learning is implemented in this research. RNN is used in this study to calculate the SILS among the students English language skills of the online schools.

11.3 METHODS AND MATERIALS

An academic instructor association is required to smooth out the progress of the student who might still triumph over the coming to a standstill of his problem-solving skills. The student's abilities can be represented as a manufactured set and by exchanging of a few words the object arrangement that is fundamental to problem solving can be accomplished. Then, the training given in the problem-solving environment must be fulfilled. The system should strive to spread an abstract sense of the challenge of resolving knowledge that occupies less memory load, and instant feedback on errors made is provided. Adjust the grain size of teaching with tutoring and, then, smooth out the progress of successive approximation to the target skill. In providing for the nonattendance of face-to-face communication lags, the students' tutoring inspiration in the aspects of web-based interactive tutoring system is given a push forward.

For these reasons, and for the proper utilization of the communication stuff, the boundary between teacher and student and self-adaptableness of students, as well as improve the communication between them, a new web-based standard intelligent tutoring system is introduced. Then, in order to strengthen the tutoring motivations of students, and based on their results, to perk up teaching priorities, a web-based Classter tool is used. The amazing advancement of artificial intelligence in e-learning brings the guarantee of its fulfilling the increasing demand to acquire multi-disciplinary, multilingual knowledge and skills to the most. The school's staff and students must put pressure on the academicians to ensure transfer of adequate quality knowledge and skills. This research discusses the main components of Standard Intelligent Tutoring System (SILS). The system provides faculty members with a simple and detailed method for tracking the success of students by browsing a range of intelligent systems reports. Figure 11.1 shows the basic intelligent tutoring system.

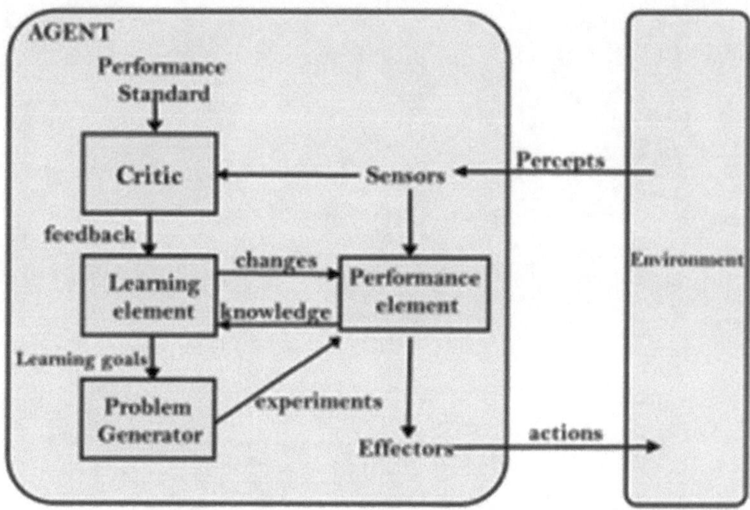

FIGURE 11.1 The basic intelligent tutor system.

11.3.1 Standard Intelligent Learning System

In the absence of a face-to-face interface, the communication lags between acquiring knowledge and its validation in an online learning system is compensated by providing the students tutoring inspiration in the form of online interaction. In this research, the SILS develops new powerful features which are fundamentally new in the conventional online examination simulator. In this paper, 665 students of matriculation school from the Bihar state are taken into account. The tutoring trends are analyzed and directed by the RNN Network in order to improve performance. Then, this network sets the questions to test the student without human intervention by taking into account the student's performance. To a certain extent than at random, a set of questions to let the examinations roll is established via online simulator. SILS is an extra open-ended system that is capable of being utilized in every field with uncomplicated configuration and modification.

Classter is a cloud-based student management system that includes all of the essential functions to run an educational institution efficiently. Classter allows students and schools to gain full access to School Management System (SMS) and Tutoring Management System (LMS) features. To unite operations, data, and employees, Classter is leveraging over 100 educational institutions their various educational processes (Figure 11.2).

Classter is a powerful modular web-based School Management System that offers complete automation of all operations of your academic institution. It combines administration, tutoring, teaching and assessment, all in a single platform. Classter allows for reasonably simple handling of applicant management, enrollment, classes, subjects, curricula and educational plans. Classter supplies web portals to any end-user type (applicant, student, lecturer or teacher, parent, employee, and alumni). It helps institutes to collect, manage and analyze student

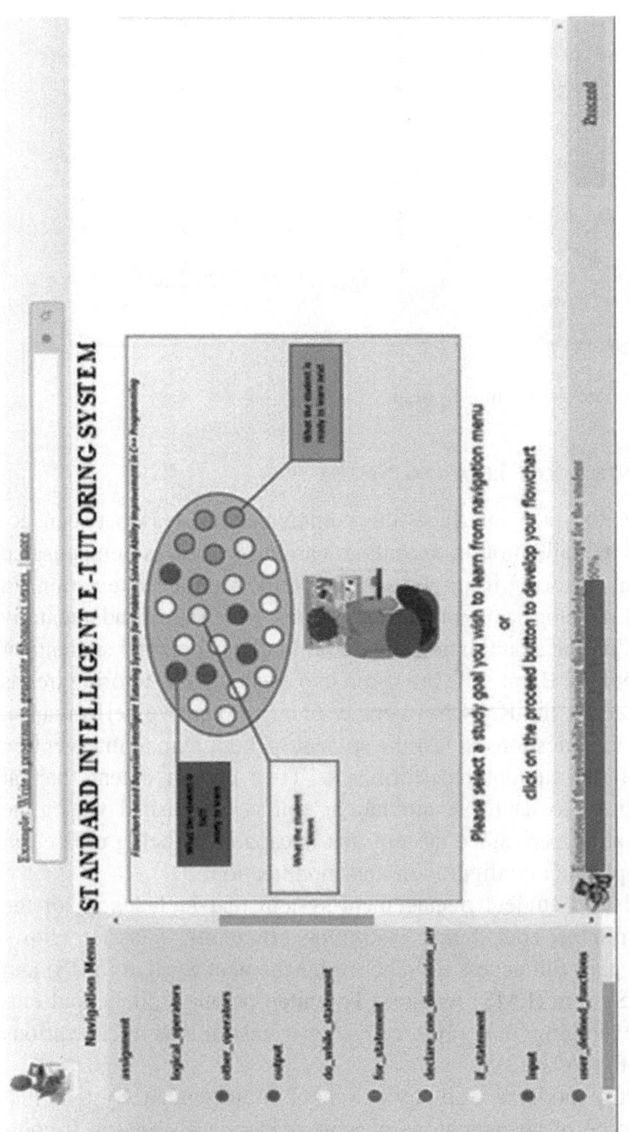

FIGURE 11.2 Standard Intelligent System.

information, access their data and monitor their progress, behavior and achievements. So this online teaching model is very useful for the students' performance calculation based on their online teaching periods. Classter allows exam schedule creation and management and storage of course content and resources. Also it may have a capacity to do the management of attendance, assessments maintenance, marking and grade reports. It provides an integrated educational C.R.M., a communication gateway, and a solution for event and meetings management. It enables school buses and transportation management, online surveys, monitoring of medical data, and much more.

In the area of education technology, Classter is a globally recognized technology that offers a single system of Cloud-based student information system (SIS), School Management System (SMS) and tutoring management system (LMS). An educational institution can use this complete and modular information management solution that incorporates end-to-end solutions. Microsoft Office 365, Google G-Suite, and other 3rd party systems such as ERP and government databases to messaging and business intelligence tools are fully integrated into the system. Classter is a comprehensive school management system that uses a single platform to deliver an integrated approach to education. Any school, from the K-12 to colleges and universities, can use the end-to-end Classter system to meet all its needs for school management.

In this research, the students online teaching states of clatter-based tutoring are taken into consideration because it is a best school-based Management System for organizing schools in its own environment. Also it is fully configurable and customizable. Clatter provides professional services and support to implement your specific requirements and adapt Classter to your needs. It simplifies roll-out and minimizes required training. It provides a personalized, user-friendly web interface for applicants, students, lecturers or teachers, employees, and parents. Classter helps management to easily reshape the entire operational process of an academic institution. It offers a secure and reliable GDPR-compliant environment. It scales easily to thousands of students and hundreds of institutions. It is supported fully by a team of experts and consultants. It is a cloud solution, but a really functionally rich one.

11.4 IMPLEMENTATION

In this paper, 665 students of matriculation school from the Bihar state are taken. These students are studying their English subjects through the web-based online teaching portal Clatter. In this paper, the student's performances of web-based online teaching portal Clatter are calculated with the clatters performance analyzer. Based on the performance of the clatter analysis results, there are three sets of students filtered. The students are separated by bad, good, and excellent. Then, based on these student's performance levels, there are three sets of questions generated. The proposed SILS develops new dominant features which are primarily innovative in the conventional online examination simulator. The RNN algorithms simulator adapts the tutoring patterns and gives any questions relevant based on the student's performances. Thereafter, this network automatically provides the questions without staff's knowledge only on the student's e-examination performance. The random set of questions smooth out the progress of the student based on the examinations which are wide-ranging and set up in online simulator. SILS is an additional helpful tool to school pedagogy to utilize in every field with less-than-par organization of school and that needs alteration. Figure 11.3 is a data-flow diagram that shows how the student interacts with the system through the user interface and receives tailored instructions after teaching strategies estimate student knowledge.

The proposed structure is a multi-agent-based model in which the used agents interact with everyone to form a unit. Each and every agent is significant in this model and they accept inputs as well as offer outputs to each of the others. Every agent has some precise responsibility as well as it's responsible to give output for the next agent to proceed. These intelligent agents are constructing units of learning to develop the information in a sequence acknowledging user interaction with the user interface.

11.4.1 Student Enrollment

Students who are enrolled with their details to the website have their details stored as information in sequence concerning the user. Initially, the user registration is needful for their assessment submission; otherwise, their enrollment will not be completed by the system. First time registers are enforced by system to provide the delicate details kept on the system. Then, the website will utilize the data and get the information rolling to the next information in sequence. At the time of registration, the users must enter their full name, date of birth, university, username and password kept on the system. The website resolves the determined values of new user as level 1 student i.e., beginner. This stage is on the other hand not permanent as the user level possibly will vary depending on their development. There are three levels of students recognized in this method: level 1 or beginner stage, level 2 or

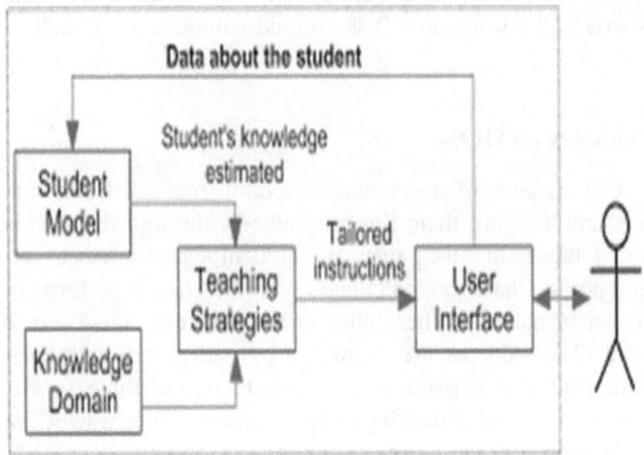

FIGURE 11.3 Student performance data collection.

intermediate stage and level 3 or advance level. The student may possibly exist in any level as well as their level based on their progress produced by the system. In this paper, 665 students of matriculation school from the Bihar state are taken. These students are being tutored their English subjects all through the web-based online teaching portal Clatter.

11.4.2 Standard Intelligent Learning System

Classter is a prevailing modern web-based school student tutoring management system that offers the whole academic institute on a computerization operation mode. It combines the manpower processes like administration, tutoring, teaching and assessment, all on a single platform. Given Classter allows easy management of candidates, admission and registration, classes, courses and English subjects, curriculums and educational plans, therefore this new online teaching model is very helpful for the student's performance evaluation on their online teaching periods. Classter allows exam schedule creation and management, and storage of course content and resources. Classter is a global, groundbreaking tool in the field of Education Technology, offering an all-in-one Cloud-based SIS, LMS. This research area offers an end-to-end and modern information management solution that can be used by any educational institution. It is completely incorporated with Office 365, Google G-Suite databases to SMS services and BI tools. Classter is a back-to-back school information system that can be used by any university, from K-12 to schools and universities, to cover your entire school's process management needs. In this paper, a student's online teaching state of Classter-based tutoring are taken because it is a best school based Management System for university students. Also it is entirely configurable and alterable. Classter supports a custom-made, comprehensible web interface for applicants, students, lecturers or teachers, employees, and parents.

11.4.3 Evaluation System

The evaluator system works to calculate the student's performance all through the e-tutoring progression. At each and every instance, the student must answer the questions posed; then, based on the answers, the RNN algorithms capture the retrieved data and evaluate the student's recital of his progress. The RNN gives the impression of being aware of the student's weakness and strength, and give information for the tutoring agent to decide the set of question to provide to the students. An RNN's responsibility is to accurately identify the student's mistake and formulate the next question to let them know their mistake. Or, the RNN approves the student's answers to the question before going to the next question. The RNN is also documenting the time taken by the student to answer a single question and for completing all the questions Figure 11.4 shows an RNN network where student's answer to a question is input in its more than one possibilities which interact with radial basis functions in the hidden layer to output an approval to progress to the next question or correct the error so that the student knows his mistake.

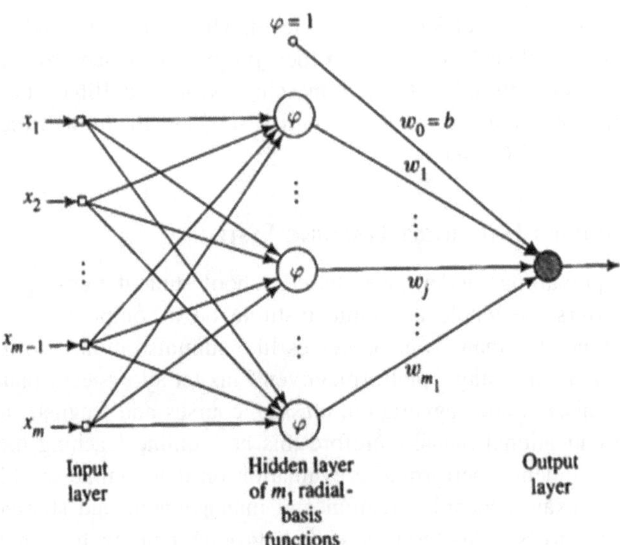

FIGURE 11.4 Basic RNN Network.

The RNN provides the students strength and weakness to answering the questions for supplementary development. Students who fail to answer a question in the approved manner had to accomplish alteration through the hints or notes provided by the user interface. Each and every user action in answering the questions will be recorded by the RNN to provide useful information to the Classter. RNN-based expert system is the alternator to simulate the human skills. In these cases, the teachers can identify the variety of students through expert field. This RNN expert structure contains the values from the e-tutoring student's models and evaluates their capability. The information gained from the e-tutoring student's models is stored in layers bank and the set of regulations are completed to maintain the information. In this work, the RNN expert model is proposed to help the online teacher to develop the collected data commencing the student's behavior. In order to provide and attain the students' level, question-based results is the most important task in this work. There are three set-off questions that are developed for the students according to average students, below average and excellent students.

11.5 RESULT AND DISCUSSION

The proposed method is a multi-agent oriented model inside the worn agents that interact with everyone to form a single unit. In this work, there are three agents to accept inputs as well as offer outputs to each one. The first one is a data provider which provides the students data and their enrollments. The Classter-based web server online teaching model is used. It has some precise responsibility as well as being responsible to give output to RNN to proceed. These RNN intelligent agents are constructed to develop the students data based on their performance sequence acknowledged from user interaction with the arrangement interface. Finally the questions are distributed by the AI-based RNN SILS.

11.5.1 Classter Student Performance Assessment

In this paper, the Classter Standard Intelligent Tutoring System is the main component of web-based online teaching process. It offers an effortless and excellent way to enroll the students who are required to learn through online teaching processes. To facilitate faculty members, the system generates numerous insightful intelligent reports to keep an eye on the performance of the students. Students who are willing to get enrolled, their details are provided to the clatter online teaching site for their enrollment process. The clatter site will detail the information in sequence concerning the students. At the first time, the student's registration is needed for their assessment submission through the system to happen. First time registers are force to provide the delicate details keen on the system. Then the website will utilize the data and get the information ahead of the next information. At the time of registration, the users are needed to enter in their full name, date of birth, university, username and password kept on the system. The website resolves the determined values of new user as level 1 student i.e., beginner. This stage is on the other hand is not permanent as the user level possibly will vary depending on their development. There are three level of student recognized: level 1 or beginner stage, level 2 or intermediate stage and level 3 or advance level. The student may possibly exist in any level as well as their level based on their progress produced by the system. In this paper, 665 students of matriculation school from the Bihar state are taken into consideration. These students are selected from their tutoring their English subjects through the web-based online teaching portal Classter (Figure 11.5).

After the registration process, the students can select their online teaching classes, provide the delicate details kept on the system. Then, the website will provide the details about the subject with ILS curriculum. Later, the site will provide the number of hours needed for that course and the books and notes and the face-to-face seminars to be conducted. Thereafter, it will maintain the day-wise attendance of the enrolled students and their assessments based on the students' performance. The website resolves the determined values of the beginner (Figure 11.6).

In the beginner stage, the new student to online teaching process must take some time to show progress. The students level possibility will vary depending on their development. There are three levels of students recognized in this method: level 1 or beginner stage, level 2 or intermediate stage and level 3 or advance level. The student may possibly exist in any level as well as their level based on their progress produced by the system. Figure 11.7 shows the dashboard of the school administrator with school's number of students enrolled, groups and teachers employed, and other quick action buttons like create a student/teacher, manage attendance, messages and user accounts.

After the completion of the allotted hours, the student's performances are arrayed and stored in the web-based server. From these figures, we can conclude that the Classter server can provide the details of the individual students as well as the

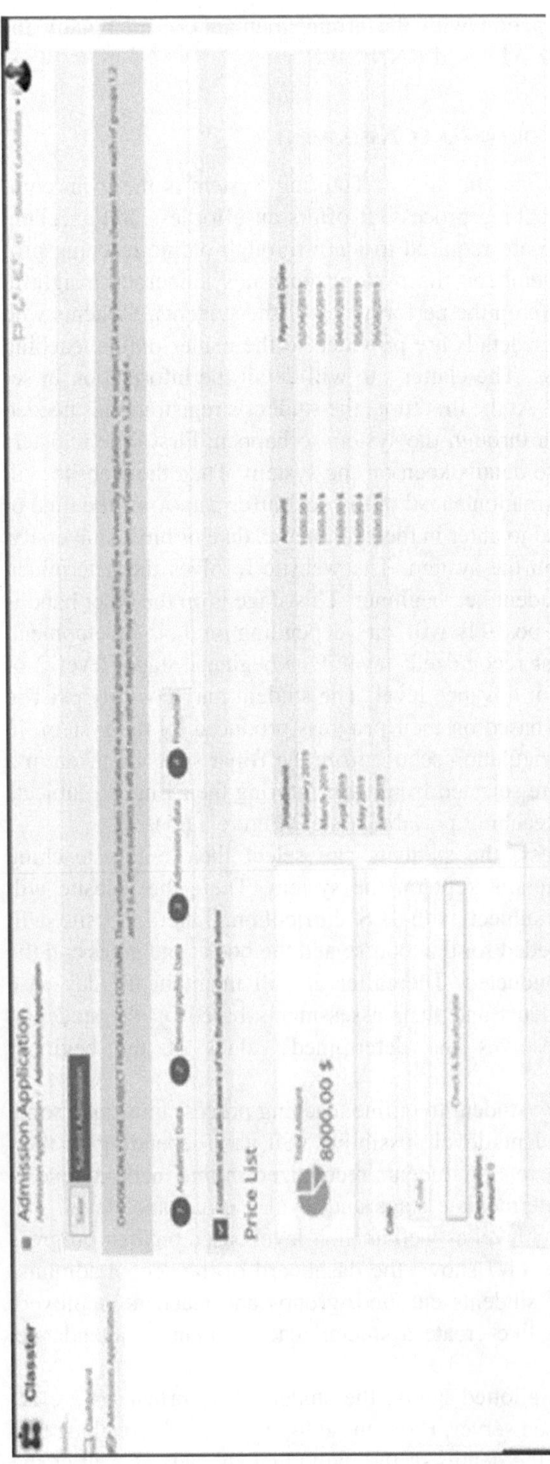

FIGURE 11.5 Shows the details of Classter Student Enrollment page.

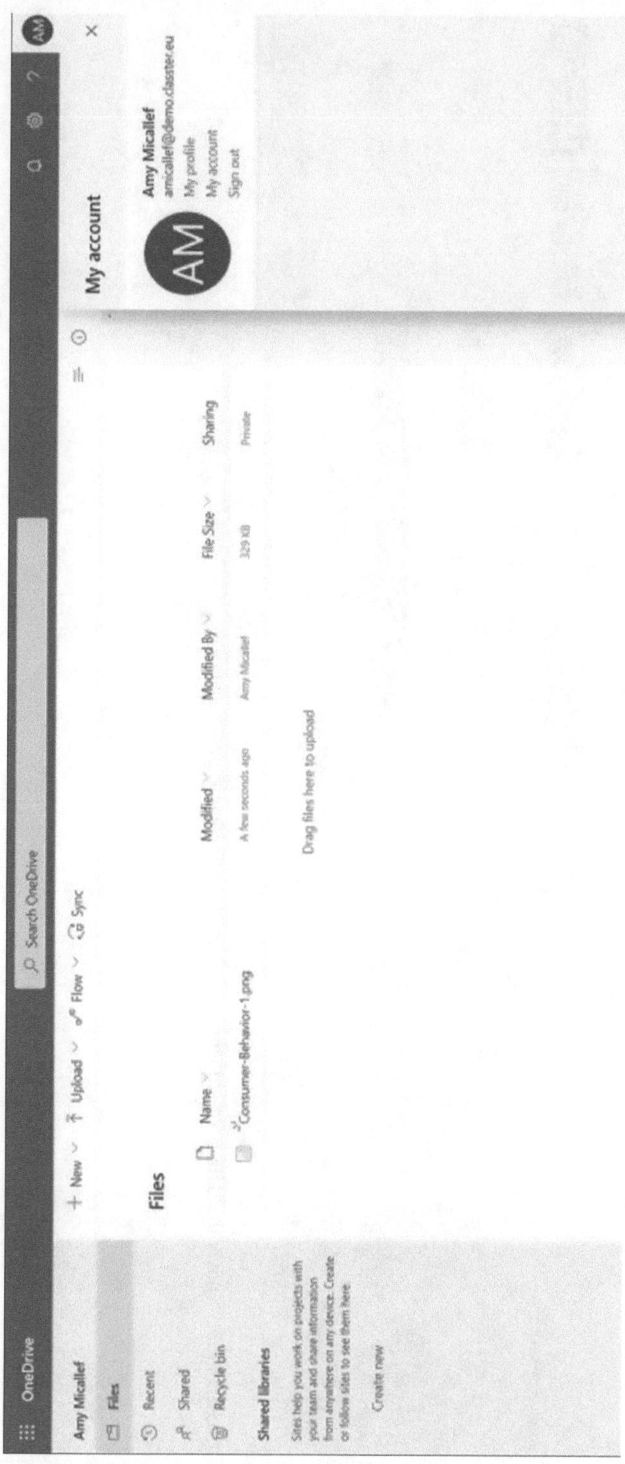

FIGURE 11.6 Shows the individual student's profile who is enrolled in Classter online teaching portal.

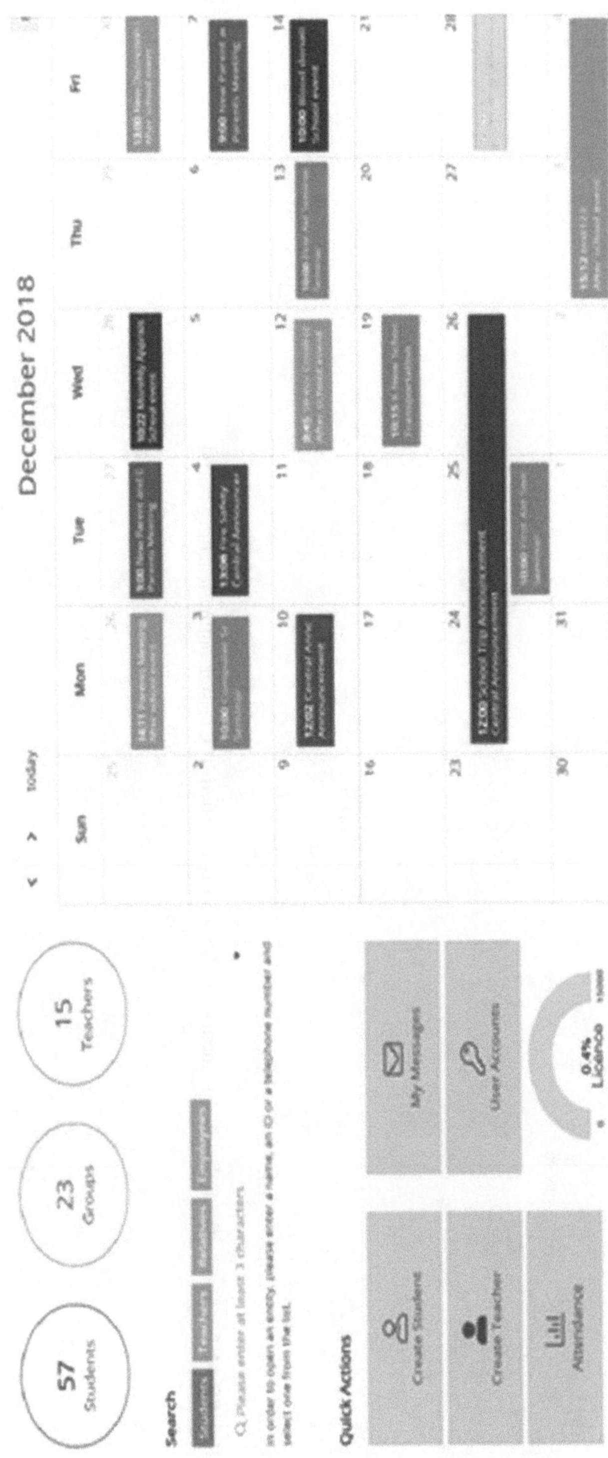

FIGURE 11.7 Shows the details of students based on their academic performance.

whole class. Also, these web servers will provide the details of the whole university's students. The best part in this is the students and staff only login through the web page. So, nobody can steal the university's student's details and the database. Then, the assessments are allocated to the students based on the previous batch student's performance, particularly, in English subjects. The beginner stage students in online teaching process are not easy with progress improvement. So, the web server checks the student's performance based on the date of joining and students level possibility. The student's attendance and their assessment tests will vary depending on their progress. There are three levels of students in this method: the student with low attendance and low assessment are categorized by level 1 or beginner stage; the level 2 student, in comparison to level 1 student, who are high in attendance and in assessment are in intermediate stage. And, the students with comparatively high attendance and high assessment are in level 3 or advance level. The student may possibly exist in any levels and their level is based on their progress produced by the system. Figure 11.8 depicts the student performance via graphs based on their levels.

11.5.2 RNN Network

RNN Network is an exacting kind of neural network model. Within this paper, a non-linear neural network model based classifier is designed. Where normally researchers used the Standard Intelligent Tutoring System as artificial neural networks, here the RNN algorithms artificial intelligent tool improves the faculty members monitoring of students. Then, performances of the students are calculated by the intelligent reports generated by the network. The online teaching-based Tutoring System develops so many features which are fundamentally most recent in the predictable online examination simulator. Every neuron in a multilayer model took the weighted computation of their input value. All the input values are multiplied with coefficients; in addition, the resultant values are added collectively. A single multilayer model neuron is an uncomplicated linear classifier; besides, the composite non-linear classifier is building with the neuron combination (Figure 11.9).

RNN performs categorization through measuring the input comparison to the values which are present in the training set. Every RNN Network neuron supplies a prototype that might be the immediately individual value as of the training set. At the time of new input classification, every neuron generates Euclidean space between the input and their prototype. There are three types of prototypes for question distribution (Figure 11.10).

If the input values are intimately resemble means it resemble the class A prototype; or else class B prototypes. If both the conditions are not satisfied it is class C. In this class-based Radial Basic Network Function adapts the tutoring pattern and provides feedback based on the student's performance criterion. Then RNN Network based on their classes and the three case-based set of questions are provided without human interference. Classes and the three case-based set of questions are framed by the student's performance. To provide a specific amount of questions

FIGURE 11.8 Shows the Classter-produced performance graph of the students based on their level marks.

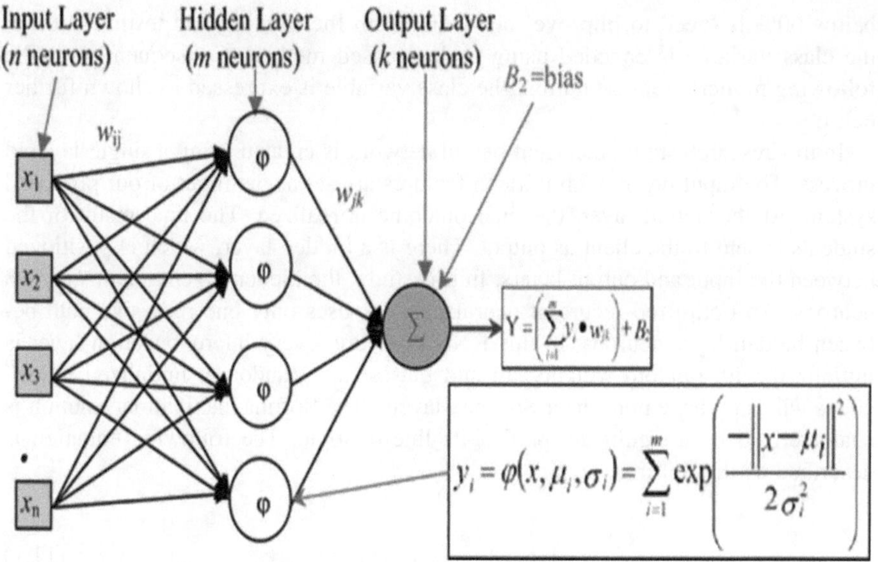

FIGURE 11.9 RNN Network.

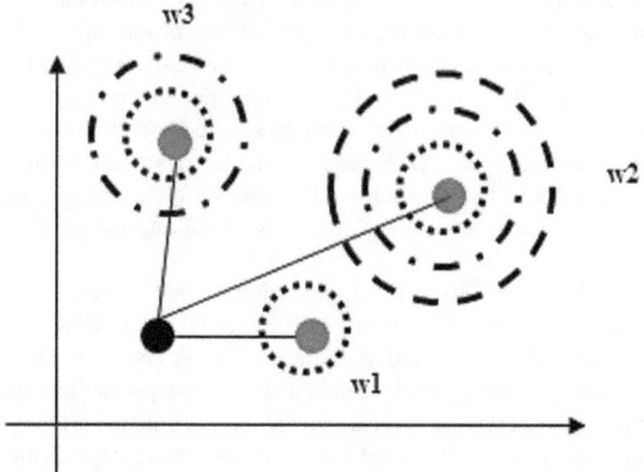

FIGURE 11.10 Shows the weight values for nearest neighbor calculation process.

rather than at random set of questions even improve the progress of the examinations.

The patch formations are based on the online test marks of the students. At the same time, it's the SILS skill test mark in range of above 80% is mentioned as 'very good' batch. Test mark in range of 80%–60% is 'good' class; and mark below 60% is 'need to improve' class. For both the training and testing dataset, the class variable is encoded using a class-based method. Consequently, in the following numeric representation, the class variable is expressed as shown further below.

In this research study, recurrent neural network is created using a single-layered process. The input layers with thirteen features are set as the input of our proposed system. At the output layer, the final outcome is realized. The final result of the students is sent to the client as output. There is a hidden layer, which is positioned between the input and output layers. In this study, the hidden layer contains of 108 neurons. An optimized recurrent neural network uses only one recursive path between hidden layer neurons. In this RNN structure, every interconnection layer is initialized with random weights (v) and biases (a_j). Randomly initialized hidden nodes will only have non-linear Softmax layers. The Softmax activation function is characterized by a significant part of its line of logic. The following equation is described as the Softmax layer:

$$f(b) = \frac{e^{ki}}{\sum e^{bj}} \max(1, v) \tag{11.1}$$

where, 'v' is the input to an input neuron, v = 0, 1, …, kj.

Softmax layer handle multiple class layers and produce only one class in other activation function normalizing the output for each class between 0, 1 and 2. The classes are separate by their sum and their probability of the input values arranged in a specific class. The output neurons are characteristically separated into Softmax and is worn only for the output layer. Recurrent neural networks are needed to classify input into multiple categories. Teachers have been shown to be optimistic about emerging technologies. SILS tends to make it smoother and less nerve-wracking for students to use computers. This can be observed as a result of how teachers are ahead of the curve when it comes to making use of IT in education today.

Then the validation process with the precision, recall, F-measure and accuracy values are evaluated. F1 = TP/TP + FP; TPR = TP/TP + FN; FPR = 2 * (Recall * Precision)/(Recall + Precision) and Accuracy = TP + TN/TP + FP + FN + TN. Precision is defined as the properly calculated positive predictions to the totality calculated positive observation. Recall is the proportion of properly calculated positive predictions to the totality calculated dependent classes. F-measure is the weighted standard of Precision and Recall. Consequently F-measure is obtained by taking both false positives and false negatives.

Table 11.1 shows the output result normalization results. From the table, the proposed RNN error control model is attained as a best outcome in all the evaluation criterions. Tutors are required to learn appropriate things for their tutoring supports. The required skill adaptations will increase their current technology adaptations and their tutorial practice improvement. If they are familiar with the ICT skills, they cannot stand in back of others.

TABLE 11.1
Output result normalization result

Methods	F1 (%)	TPR (%)	FPR (%)	Accuracy (%)
Traditional RNN	51.24	65.52	54.43	64.53
RNN-gated	62.42	63.42	65.82	65.05
RNN-multiplicative	63.25	64.31	66.31	65.90
Proposed RNN-up and down sampling	75.11	78.19	76.51	78.24

11.6 CONCLUSION

The online teaching based Standard Intelligent Tutoring System develops new powerful features which are fundamentally most up-to-date in the predictable online examination simulator. The RNN algorithm adapts the tutoring patterns and gives feedback based on the student's performance decisive factor. Then, RNN Network sets the questions with no human intrusion for the student's presentation. To an unambiguous amount than at random, a set of questions to even the progress of the examinations which are wide-ranging are found in the online simulator. SILS is an extra benevolent that is competent to be utilized in every field with uncomplicated pattern and modification. In this work, the case-based intelligent assessment arrangement is used. It is an alternate for translating professionals information to regulations. In this model, the student's data and information are corresponding to their subject field information. In this work, the cases are developed to pose relevant questions to the students. The cases are likely to be the understanding of demonstration for the personal belongings of students administer to articulate their study and suggestion or explanation, instruction, and strategy of students. All the levels are in a sequence and all are prearranged cases with the purpose of resulting. The procedure of constructing the cases involve collecting data of a number of students. All the composed information are verified by the server. All the information have been equipped in typical appearance of cases and stored in the folder for supplementary use in the system.

REFERENCES

1. Kumar, T. (2020). Impact of motivation and group cohesion in EFL classrooms at Prince Sattam Bin Abdulaziz University, KSA. *Asian EFL J.*, 27(4.4), 116–131.
2. Koutina, M., & Kermanidis, K.L. (2011). Predicting postgraduate students' performance using machine learning techniques, in *Artificial Intelligence Applications and Innovations*, Springer, Berlin, Heidelberg, 159–168.
3. Alotaibi, S.S., & Kumar, T. (2019). Promoting teaching and learning performance in mathematics classroom through e-learning. *Opción*, Año 35, Especial No. 19, 2363–2378.
4. Beatson, R.K., & Newsam, G.N. (1992). Fast evaluation of radial basis functions: *I. Computers Math.*, 24(12), 7–19.

5. Angros, R., Johnson, W.L., Rickel, J., & Scholer, A. (2002). Tutoring domain knowledge for teaching procedural skills, in Gini, M., Ishida, T., Castelfranchi, C., & Johnson, W.L. (eds.) *Autonomous Agents and Multi-agent Systems*, ACM, New York, 1372–1378.
6. Arroyo, I., Cooper, D.G., Burleson, W., Woolf, B., Muldner, K., & Christopherson, R. (2009). Emotion sensors go to school, in Dimitrova, V., Mizoguchi, R., du Boulay, B., & Graesser, A. (eds.) Proceedings of 14th International Conference on Artificial Intelligence in Education, pp. 41–48.
7. Baghaei, N., Mitrovic, A., & Irvin, W. (2006). Problem-solving support in a constraint-based intelligent tutoring system for UML. *Technol. Instr. Cogn. Learn.*, 4(2), 113–137.
8. Baghaei, N., Mitrovic, A., & Irwin, W. (2007). Supporting collaborative tutoring and problem-solving in a constraint-based CSCL environment for UML class diagrams. *Comput. Support. Collab. Learn.*, 2(2–3), 159–190.
9. Barros, B., & Verdejo, M.F. (2000). Analysing student interaction processes in order to improve collaboration: The DEGREE approach. *Artif. Intell. Educ.*, 11, 221–241.
10. Barrow, D., Mitrovic, A., Ohlsson, S., Grimley, M. et al. (2008). Assessing the impact of positive feedback inconstraint-based ILSs, in Woolf, B. (ed.) Proceedings of 9th International Conference on ILS 2008, LCNS 5091, pp. 250–259. Springer-Verlag, Heidelberg.
11. Billingsley, W., & Robinson, P. (2005). Towards an intelligent online book for discrete mathematics, in Proceedings of International Conference Active Media Technology, pp. 291–296.
12. Nkambou, R., Mizoguchi, R., & Bourdeau, J. (2010). *Advances in Interactive Tutoring System*, Springer, Heidelberg.
13. Ohlsson, S. (1992). Constraint-based student modeling. *Artif. Intell. Educ.*, 3(4), 429–447.
14. Mitrovic, A., & Ohlsson, S. (2006). Constraint-based knowledge representation for individualized instruction. *Comput. Sci. Inf. Syst.*, 3(1), 1–22.
15. Ohlsson, S., & Mitrovic, A. (2007). Fidelity and efficiency of knowledge representations for interactive tutoring system. *Technol. Instr. Cogn. Tutoring*, 5(2), 101–132.
16. Dahl, G.E., Yu, D., Deng, L., & Acero, A. (2012). Context-dependent pre-trained deep neural networks for large-vocabulary speech recognization. *IEEE Trans. Audio Speech Lang. Process.*, 20(1), 30–42.
17. Collobert, J.W. (2008). A unified architecture for natural language processing: Deep neural networks with multitask learning, in Proceedings of the 25th International Conference on Machine Learning, pp. 160–167.
18. Jia, Y., Shelhamer, E., Donahue, J., Karayev, S., Long, J., Girshick, R., Guadarrama, S., & Darrell, T. (2014). Convolution architecture for fast feature embedding, in Proceedings of the 22nd ACM International Conference on Multimedia, pp. 675–678.
19. Kyndt, E., Musso, M., Cascallar, E., & Dochy, F. (2011). Predicting academic performance in higher education: Role of cognitive, learning and motivation, in Earli Conference.
20. Livieris, et al. (2012). Predicting students' performance using artificial neural networks, in 8th Pan-Hellenic Conference with International Participation Information and Communication Technologies, pp. 312–328.
21. De Mulder, W., Bethard, S., & Moens, M.-F. (2015). A survey on the application of recurrent neural networks to statistical language modeling. *Comput. Speech Lang.*, 30(1), 61–98.

12 Fusion of Datamining and Artificial Intelligence in Prediction of Hazardous Road Accidents

Bilal Khan[1], *Asif Hasan*[2,3], *Digvijay Pandey*[4], *Randy Joy M. Ventayen*[5], *Binay Kumar Pandey*[6], *and Gadee Gowwrii*[7]

[1]Department of Computer Science, University of Bradford
[2,3]Assistant Professor, Department of Psychology, Aligarh Muslim University
[4]Department of Technical Education, IET, Dr. A.P.J. Abdul Kalam Technical University
[5]Dean (CBPA), Pangasinan University, Philippines
[6]Assitant Professor, Department of IT, College of Technology, Govind Ballabh Pant University of Agriculture and Technology
[7]MSc (Statistics), Osmania University

CONTENTS

DOI: 10.1201/9781003194415-12

12.1 INTRODUCTION

The growing population and urbanization are causing hectic and complicated transportation system. The effect of overland transportation can be externalized sometimes, as accidents. The past data shows heavy loss of human life due to sudden accidents. The structure of road and traffic is increased with a new vehicle every year. The new number of vehicles is up to one million every year. The record shows, each year, around 1.2 million people have died due to road accidents and around 50 million have been injured. Therefore, an effective prevention and solution system needs to be put in place to minimize or stop road accidents for the benefit of human lives. For safety in the road against accidents, prediction can be helpful. This prediction is based on data collection. Hence, prediction for an accident in the road is considered to be a critical part of the safety program.

The most effective way to minimize road accidents is the use of a subsidiary discipline of artificial intelligence which is called machine learning. The use of machine learning technique is boundless. Machine Learning uses different functional methods for granting legitimate and reliable decisions with the help of experiences. The best part of machine learning is that it can import information from the source of data and use this in a statistical method. Thus, creating a machine learning-based application can be very effective in predicting road accidents with help from data.

Data mining is the answer to understand road accident when the data stored, analyzed, compare with other studies. Given the space and time complexity of data quickly increases without an advanced data management system, this disadvantage in handling data is overcome with data mining. Data mining that searches out patterns in haphazard, billowing datasets makes this technique more compact and suitable for exploitation in this field. The pattern recognition and prediction of an incident are its main features, whereas in a road accident, the data and the other cases related to this event increase day by day, and for this, the data collection becomes huge. The data mining emerging techniques can handle this data perfectly. If by employing this technique, the situation of the states will improve and there is a fall in road accidents, then the success of data mining will be in its saving human lives.

Prediction of road accident before it can happen will save many lives by avoiding accidents. There are certain models for this. Artificial Intelligence (AI) can detect the patterns of unsafe driving of vehicles in the road for prediction purposes, such as detection and prediction of unsafe driving and crash.

12.2 RELATED WORKS ON PREDICTION OF ROAD ACCIDENTS

The data mining process is helpful to predict the future by finding out how accidents happen on roads and how to avoid accidents on road. The road accident is one of the main problems at the global level. Most of the developing countries have been suffering from this critical problem due to the unavailability of modern technical support and proper infrastructure. The data mining method is used to extract the summary of all kinds of previous road accident cases and arrange raw data. It then processes the data to make a proper database for future use. Data Mining contains different kinds of data and analyzes them with proper resources. The modern technique is also identified as "KDD or Knowledge Discovery in Data". The system is helpful to improve the safety structure of roads to provide better protection of the common people's lives. In this phase, some of the related case records have been highlighted to overcome the current situation of hazardous road accidents and predict the possibility to reduce the road accidents.

In the global platform every year, approximately more than 1.40 million common citizens are affected by the road accident. On the other hand, 50 million people are facing small injuries with many disabilities. In the present condition, the study helps understand the local traffic management apparatus to prevent accidents on road. India is a major populous country which is facing the road-related problems every day. Road accidents are a common factor to increase the mortality rate in India. Iran and Iraq have highly been affected by road accident-related deaths and major casualties due to the proper application of infrastructure and modern techniques to prevent road casualties. The last three-year data analysis indicates the average common citizen deaths due to road accidents.

The illegal activities on roads and avoidance of traffic rules have affected the common citizen's life [1]. Since more than 25,000 people are affected by road accidents in this nation, the Indian government has introduced different kinds of traffic rules to ensure public safety on the road. Given the fact that rash-and-drunken driving is one of the major reasons to increase road accidents, therefore accident-related information is helpful to prevent road accidents, even as security of travelers will be taken up by the local government authority. In many of the developing countries of the world, there are hired professionals who have skilled people to measure the possible risk to human life on the road. The safety and security of life have always been important to measure the road accidents taking lives or causing injury, and improvisation of the traffic rules. The Information Ministry of Iraq has introduced modern technology to improve the road traffic system. The main target of this implementation is to provide fewer casualties, take proper safety policies to recover from the critical conditions and prevent rising mortality from road accidents [2].

India is one of the quick developing countries globally. The nation is not only dealing with the traffic system but also fighting with a huge population and consequent food crisis. The poverty level is increasing day by day with land degradation. The possible data analysis has indicated that more than 45,000 speed violation cases recorded in Mumbai city. In 2017, Kolkata as the most famous city of India as far as road accidents are concerned faced huge casualties from bad traffic management, violation of traffic rules, illegal construction on the roadsides, etc. The data analysis highlighted the mortality rate increases more than 19 times from where it stood previously. The majority of the people suggested installing modern techniques to reduce road casualties. The suggestion of the government to improve road traffic infrastructure has provided modem speed detectors tools to analyze the speed of vehicles, install CCTV cameras in all-important connectors, provide night control room support for quick recovery, to name a few, and install "Automatic number plate recognition system" to locate the criminal [3].

Bangladesh one of the developing countries of the world faces poverty and illegal migration. A major development here has been to deflect poverty by the proper implementation of government laws to reduce these critical conditions. Political issues have deflected the main motive of safety and security for the common citizen. The most populous region of Dhaka has many illegal vehicles captured from the road, and the commoners are daily suffering from the traffic system. The common people avoid traffic rules to reach the target area within time. Due to this reason, many casualties have occurred here, the mortality rate also increased for violation of the traffic rules. More than 1025 people are affected by vehicle accident [4]. The common citizens have tried to maintain the traffic rules to avoid the casualties but the illegal vehicles, rash driving, avoiding the traffic rules affected the lives of the common people. The global ratio has indicated the adaptable vehicles also were the cause behind road accidents.

Most of the cases have revealed the main reason behind road accidents from lack of proper knowledge of traffic rules. The drunken driving, avoidance of the red light, excessive speed, not using the safety-belt and helmets are indicated as the main causes of fatalities on the road. The Hong Kong government has separated the transportation system to maintain the traffic rules and regulations, and to provide better safety and security to its citizens. The government has secured all data in a cloud system to measure and analyze the data for future use [5].

The globe's most popular tourist destination, Hong Kong, represents south China. Last year, more than 3000 road accident cases were reported due to avoidance of the traffic rules. The government has ensured to provide the best experience in road communication. Having introduced modern technology which is helpful to decrease the road accident [6], the government has ensured strict observance of rules and regulations to ensure security of lives on roads. The analytical report has reflected that the government of China decreased the total road casualties from that of the last year. The vehicle accident report analysis had confirmed it as only 227 people having suffered due to road accidents. The procedure of the improvement has indicated the critical conditions to be overcome and to provide life security [7]. In critical conditions, the data mining procedure has been more helpful to locate the critical problems and to justify them. The possible prediction may have a proper base on the database, which stores data on the possible factors, conditions of the roads and proper traffic management to avoid road accidents.

12.3 MOTIVATION AND PROBLEM STATEMENT

The major road accidents have occurred due to avoidance of traffic rules. The World Health Organization (WHO) has estimated the cost of total number of road accidents which increased the mortality rate in the globe. A research report has revealed the truth that more than 1.50 million people were dying due to road accidents every year, globally. The report stated the machine learning procedures which help to decrease the road accident casualties and to make prediction to overcome the critical situation. The machine learning process is used to collect data from a previous case study and through AI techniques. The collected data will help to measure the critical situation from road accidents, and to provide a database report to overcome the difficult situations. The data exploring helps to measure the problems with the machine learning process. The whole process is to use modern technology to analyze the critical data and reflect a result on datasheet with graphical representation.

The machine learning process approaches a model to analyze the problem in different ways. Thus, it is easy to take regression problems and analyze them. After the observation, the number of predictions is based on fatalities on the crash datasheet. The approach is classified as the problem in different portions and prediction depends on the acutance of the provided database. The collected data must be really taken from a previous case study, and this data helps to judge the critical situations and provide proper feedback support to prevent the conditions.

12.4 PROPOSED METHODOLOGY

The process of this work depends on data collection and data mining procedure through AI techniques. The data mining procedure helps to collect data from different resources. The critical situation analysis helps to predict the critical situation on the road accident. The machine learning procedure helps to build the accident prevention model and decrease the mortality rate in the global region. It was observed from in the different case studies, the procedure can assume the condition and is able to predict the disaster effect. There are two following datasets that choose to provide the final results in this study report.

12.5 KAGGLE AND GOVERNMENT STATISTICAL DATA

The dataset has been used to determine the record of road accidents which is published on online government websites. The possible data has been measured with Kaggle datasets. From the process of data mining, we have spotted the causes as the below-par knowledge of population and road traffic among people, and they are losing their lives consequently. Highly developed countries are maintaining their traffic system in well-developed manner with the help of technologies, where the collected data has indicated the most number of casualties due to rashness and over speed driving. The data process is reflecting the defaulters of traffic control which can affect human life. The defaulters suffer from the lack of concentration and are able of causing the major health injuries. The data has contained the proper location, time management, total casualties due to road accidents and various traffic dropouts. The project has observed the critical analysis of road accident in different regions of the globe. The process of data mining is helpful to locate the disputes.

The main cause of death and hospitalization in India is road accident. In 2016, she occupied the 8th position in the list of countries causing death from accidents around the globe. The age group of people who are losing their lives under this cause is the 15–49 years age group.

12.6 DARK SKY

The environmental factors have increased road accidents on a global scale. The previous dataset process Kaggle had contained various meteorological data information. In this assessment, the critical analysis of data mining process helps to locate the various environmental conditions which lead to road accidents in regions around the globe and in the Asiatic region. The previous process is not sufficient for data analysis. In this process, two main issues have been observed from data analysis. The two different weather conditions are mentioned below:

12.6.1 The Datasets Help to Assume the Constant Weather Conditions on the Whole Day

In the case study of Asiatic regions, various parts of the country was suffering from overpopulation, which has caused severe degradation rate of quality of life. The majority of people live in a congested area so that narrow and damaged roads lead to accidents in a manner of surety. The drainage system has comparatively affected road accidents. A majority of vehicle users avoid the traffic rules. This is one of the reasons which lead to road accidents. The proper weather conditions have not able to preside due to frequent changes in the weather factors.

12.6.2 The Environmental Factors Depend on Previous Environmental Datasets

In the basics of previous datasets, any climate or weather report can be assumed to predict the upcoming situations based on particular satellite data information. The forecast report has to be able to focus on future weather conditions. The process is helpful for any risk analysis. The ability to mimic uncertain disaster conditions in future, the weather reports have also indicated the weather changes.

The methodological path has helped to utilize the critical situation with meteorological dataset. The Dark Sky has been able to determine the special factors of environment-related causes of road accidents. The process helps to analyze the datasets and reduce the possibility of a road accident. The Dark Sky is able to provide past and future reports, and measure critical weather based on the comparison. The process is updated every half hour based on provided data resources. The process reflects the critical situation with proper datasets results.

The data mining technique Apriori and Naïve Bayes are applied to predict road accident patterns. These classifiers are specifically used because it will provide best results while working with more datasets and records. In Figure 12.1, the architecture diagram to predict road accidents is depicted. To use the data effectively, we collect data in a data repository.

The methodology used in this report predicts the kinds of accidents occurring on roads by using 61 constraints. Some of the constraints are the conditions of weather, surface of road, vehicle types and age, gender and age of a person, reference number of a vehicle.

A model is developed to predict the type of accident occurring frequently with the help of association rule. It will be done by using input as association rules for new road.

12.6.3 Apriori Algorithm for Road Accident Prediction

1. Find the support system by using scanned data of each item.
2. Frequent one item set (L1) should be generated. To generate set candidate, of k-item set, use LK-1, and join LK-1 with it.
3. The candidate k-item set should be scanned. Then, the support of each candidate k-item set should be generated.
4. Until it becomes C = Null Set, frequent item set should continue to be added.
5. Non empty subset should be generated for all items in the frequent item-set.
6. The determination of confidence will be done for each empty subset. It will be equipped with strong association rule, if confidence must not be lesser than nor not equal to specified confidence level.

12.6.4 Road Accident Analysis and Classification Using Apriori Algorithm

Consider Input dataset (A, B, C, D and E are accident types) (Table 12.1):

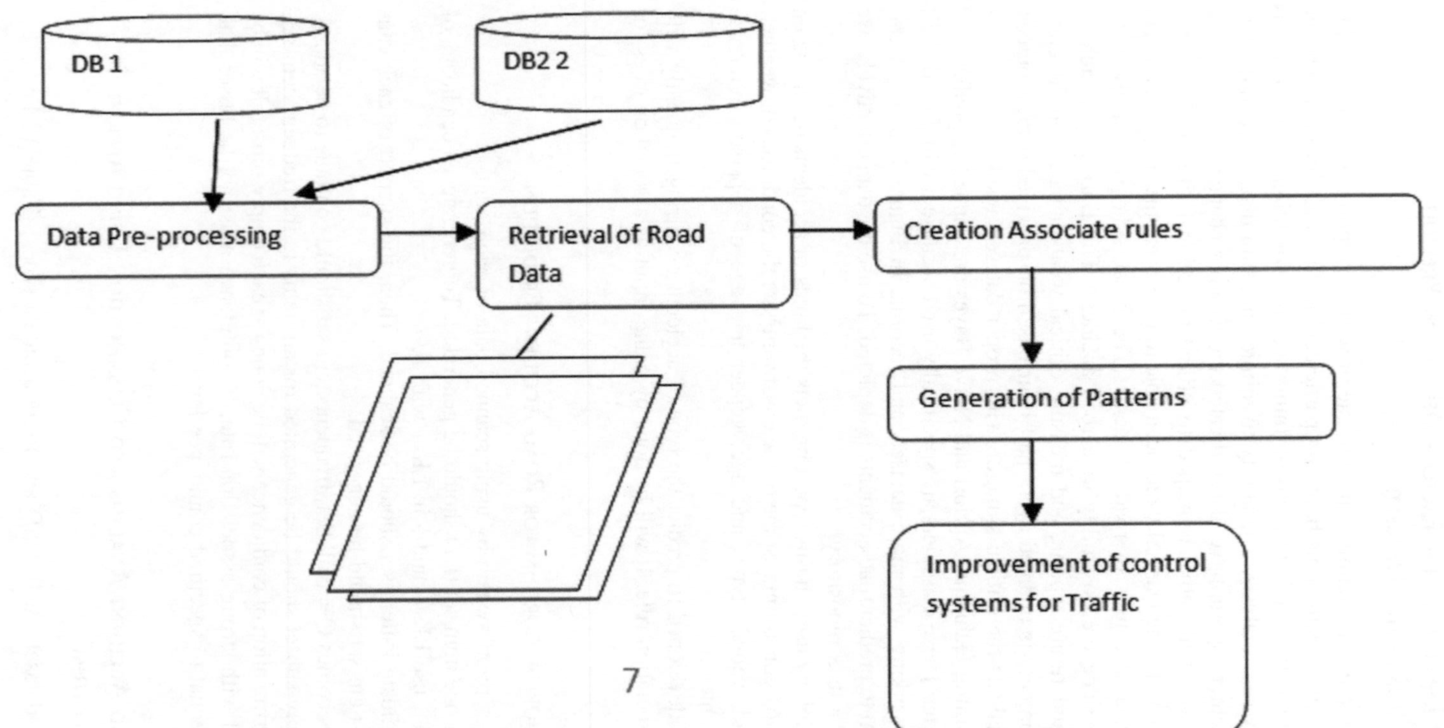

FIGURE 12.1 Database architecture data-flow diagram.

TABLE 12.1
Dataset and accident types

TID	Items
1	A, C, D
2	A, C, E
3	A, B, C, E
4	B, E

Minimum support = 50%
Minimum confidence = 80%
Item set: A, B, C, D, E

12.6.5 Strong Association Rule Mining for Road Accidents

The result which obtained is:

1. {E}<-{B}
2. {A}<-{CE}
3. {C}<-{AE}
4. {C}<-{A}
5. {A}<-{C}

To predict road accidents for the roads which are going to construct is done by classification rule. In this report, the algorithm used in the implementation of classification rule is Naïve Bayes algorithm.

12.6.6 Naïve Bayes Algorithm for Prevention of Road Accidents

1. Dataset should be scanned.
2. Each attribute value's [n, n_c, m, p] probability should be calculated.
3. The formulae given below should be applied.
4. The probabilities should be multiplied by p

$$P\left[\left(\text{value of}\frac{\text{attribute(ai)}}{\text{Subject value vj}}\right)\right] = (n\_c + mp)/(n + p) \tag{12.1}$$

where:
n = the number of training examples for which v = vj
nc = number of examples for which v = vj and a = ai
p = apriori estimate of p(aijvj)
m = the equivalent sample size

5. The classification of attribute values will be done with one of the set of classes which were predefined; this will be done after the values are compared.

12.6.7 Sample Example

Attributes (Constraints) – Speed Limit, Weather,
Pedestrian Distance [m = 3]
Subject (Accident Type) – A1, A2 [p = 1/2 = 0.5]

12.6.8 Training Dataset

The training dataset is given in Table 12.2.

Derive the accident type (A1/A2) in new road with the use of following data Speed limit is X, Pedestrian distance is R, weather is A.

$$P = [n\_c + (m * p)]/(n + m)$$

TABLE 12.3

A1	A2
X	X
P = [n_c + (m * p)]/(n + m)	P = [n_c + (m * p)]/(n + m)
n = 2, n_c = 2, m = 3, p = 0.5	n = 2, n_c = 0, m = 3, p = 0.5
p = [2 + (3 * 0.5)]/(2 + 3)	p = [0 + (3 * 0.5)]/(2 + 3)
p = 0.7	p = 0.3
A	A
P = [n_c + (m * p)]/(n + m)	P = [n_c + (m * p)]/(n + m)
n = 2, n_c = 2, m = 3, p = 0.5	n = 2, n_c = 0, m = 3, p = 0.5
p = [2 + (3 * 0.5)]/(2 + 3)	p = [0 + (3 * 0.5)]/(2 + 3)
p = 0.7	p = 0.7
R	R
P = [n_c + (m * p)]/(n + m)	P= [n_c + (m * p)]/(n + m)
n = 2, n_c = 1, m = 3, p = 0.5	n = 2, n_c = 1, m = 3, p = 0.5
p = [1 + (3 * 0.5)]/(2 + 3)	p = [1 + (3 * 0.5)]/(2 + 3)
p = 0.5	p = 0.5

A1 = 0.7 * 0.7 * 0.5
A1 = 0.1225
A2 = 0.3 * 0.3 * 0.5 * 0.5
A2 = 0.0225
Since A1 > A2

TABLE 12.2
Training dataset

Types of Road	Speed Limit (X, Y, Z)	Weather (A, B, C)	Pedestrian Distance (P, Q, R)	Accident Type
Road 1	X	A	P	A1
Road 2	X	B	Q	A1
Road 3	Y	B	P	A2
Road 4	Z	A	R	A1
Road 5	Z	C	R	A2

12.7 SOFTWARE USED FOR PREDICTION

12.7.1 JUPYTER

Jupyter is A free and open-source software. The purpose of using it is for its open standard. Also, it is used for computing services using a programming language. The benefit of using Jupyter is its capability to be used online by sharing files and lives coding. It is used for cleaning of data by simulation of numerical transformation. The models used can be presented here for the result. The obtained classified algorithm is operated and compiled in the Jupyter with the help of machine learning.

12.7.2 PYTHON

Python is basically a software with huge scope in machine learning. The level of Python is suitable for future machine learning. Having a variety of features with many active coders using it, the datasets organized into logistic regression is compiled in the Jupyter, and its subsequent algorithm is plotted in Python.

12.7.3 HTML AND CSS

The webpage development is done with the help of these softwares. The prediction result is obtained following designing UI or user interference. This UI is generated using HTML and CSS language software. This interference is indicated as a website for taking inputs from different constraints. These constraints are obtained from users in road accidents. The output is then forwarded to another program for further working on.

12.8 RESULTS AND DISCUSSION

The prediction of road accident is made using the dataset, which is organized as values and normal text. This is done for clear understanding and easily predicting numerical values. Also, using this numerical data, the calculation can be made easy.

Therefore, the dataset is organized as rows and columns from null values to numbers. The further prediction by classification algorithm is conducted on the entire dataset.

The final result of prediction by classification algorithm of datasets shows by the percentage of any particular area. The classification algorithm with less number of features in the area of research has accuracy level high with fast processing of the result. The final result of the classified regression is shown by the percentage of area by eliminating any level of errors in it.

As per the records, the following patterns are created using Naive Bayes Algorithm.

12.9 GRAPHICAL REPRESENTATION

In Table 12.3, a comparison of road accident statistics for 2017 to 2016 is performed, where observed accident range has increased to 0.4%.

12.10 ROAD CATEGORY AND ROAD FEATURES

As on 31st March 2016, the total length of roads all over India is about 56 km. The percentage of road constituted is 1.8% for National highways, 3.1% for State highways and 95% for other roads. Using this report, we observed the length and percentage given are not even. In Table 12.4, a statistics about 464,910 road accidents are recorded, National Highways (NH) occupies 30.4%, State Highways (SH) occupies 25.0% and other roads occupies 44.6%. The death recorded according to roads are 36.0% in NH, 26.9% in SH and 37.1% in other roads. 40% of road traffics are happening in National Highway(NH).

Table 12.4 represents road accidents, fatalities and injuries.

12.11 ACCIDENTS BY ROAD ENVIRONMENT

Road environment is one of the major causes for accidents. 19.4% of accidents occurred in a residential area, 12.5% in a market/commercial area. Table 12.5 shows accidents happening in Institutional area which is lower than we expected.

Table 12.6 shows 64.2% of accidents on road happened on straight roads, the remaining on roads (curved roads, pothole roads and steep grade) occupy 15.6%, while 2.5% have occurred on roads where construction work was in progress.

From Table 12.7, roads having some kind of traffic control measure implemented count for 29.5% of accidents on roads. Uncontrolled junctions cover 70.5% of accidents.

12.12 ACCIDENTS BY WEATHER CONDITION

The weather changes affect road surface condition. It makes driving riskier as fog, sun and rain complicate vision of vehicle riders. 16% of total road accidents have occurred due to weather condition in 2017 as shown in Table 12.8. Road accidents are aggregated day by day as hazardous driving is a huge concern for all the

TABLE 12.3
Road accident statistics for year 2016–17

Parameter	2016	2017	% Change over Previous Year
Count of road accidents	480,652	464,910	-3.3
Count of persons killed	150,785	147,913	-1.9
Count of persons injured	494,624	470,975	-4.8
Severity accident (persons killed per 100 accidents)	31.4	31.8	0.4*

TABLE 12.4
Road accidents fatalities and injuries data for year 2016–17

Road Category	2016 Count of Accidents	Persons Died	Persons Injured	2017 Count of Accidents	Persons Died	Persons Injured
National highways	142,359 (29.6)	52,075 (34.5)	146,286 (29.6)	141,466 (30.4)	53,181 (36.0)	142,622 (30.3)
State highways	121,655 (25.3)	42,067 (27.9)	127,470 (25.8)	116,158 (25.0)	39,812 (26.9)	119,582 (25.4)
Other roads	216,638 (45.1)	56,643 (37.6)	220,868 (44.6)	207,286 (44.6)	54,920 (37.1)	208,771 (44.3)
Total	**480,652**	**150,785**	**494,624**	**464,910**	**147,913**	**470,975**

TABLE 12.5
A comparison of accidents in various areas

Areas	Total Count of Accidents	Persons Died	Persons Injured
Residential areas	89,212 (19.2)	25,815 (17.5)	88,517 (18.8)
Institutional areas	29,670 (6.4)	9403 (6.4)	29,105 (6.2)
Market/commercial	58,166 (12.5)	17,059 (11.5)	55,376 (11.8)
Open areas	234,769 (50.5)	78,323 (53.0)	245,923 (52.2)
Others*	53,093 (11.4)	17,313 (11.7)	52,054 (11.1)
Total	**464,910**	**147,913**	**470,975**

TABLE 12.6
Number of accidents by road features

Road Features	Count of Accidents	Persons Died	Persons Injured
Straight road	298,351 (64.2)	91,203 (61.7)	302,952 (64.3)
Curved road	54,077 (11.6)	17,814 (12.0)	57,346 (12.2)
Bridge	15,514 (3.3)	5543 (3.7)	15,839 (3.4)
Culvert	11,600 (2.5)	4144 (2.8)	11,974 (2.5)
Potholes	9423 (2.0)	3597 (2.4)	8792 (1.9)
Steep grade	9124 (2.0)	3248 (2.2)	9753 (2.1)
Ongoing road works/under construction	11,822 (2.5)	4250 (2.9)	11,425 (2.4)
Others*	55,000 (11.8)	18,115 (12.2)	52,896 (11.2)
Total	**464,910**	**147,913**	**470,975**

TABLE 12.7
Number of accidents, fatalities, and injuries due to traffic control

Road Features	Count of Accidents	Persons Died	Persons Injured
Traffic light signal	16,563 (9.4)	4058 (7.5)	15,547 (8.9)
Police control	13,577 (7.7)	4114 (7.6)	12,590 (7.2)
Stop sign	11,140 (6.3)	3346 (6.2)	10,812 (6.2)
Flashing signal/blinker	10,549 (6.0)	2896 (5.4)	10,644 (6.1)
Uncontrolled	124,024 (70.5)	39,560 (73.3)	124,486 (71.5)
Total	**175,853**	**53,974**	**470,975**

TABLE 12.8
Accidents due to weather conditions

Weather Condition	Count of Accidents	Persons Died	Persons Injured
Sunny/clear	340,892 (73.3)	102,926 (69.6)	349,597 (74.2)
Rainy	44,010 (9.5)	13,142 (8.9)	46,004 (9.8)
Foggy/misty	26,982 (5.8)	11,090 (7.5)	24,828 (5.3)
Hail/sleet	3078 (0.7)	1523 (1.0)	2888 (0.6)
Others	49,948 (10.7)	19,232 (13.0)	47,658 (10.1)
Total	**464,910**	**147,913**	**470,975**

Note: Bracket values show percentage shares in the total of respective columns.

countries over the world. According to research, almost 1.4 million people die each year in accidents on road from all over the globe. The reason behind this cause for accidents can be controlled. This is not any disease that arrives without cause. From the many different reasons for death and many different kinds of threats by which human deaths happen, some cannot be controlled. But it is in the hands of humans to prevent road accidents. Hence, this can be reduced and completely demolished. Only proper education and awareness can accomplish this. There are some basic ways which will help to prevent this [8].

12.13 TYPES OF VEHICLES INVOLVED IN ROAD ACCIDENTS

The accident on national highways, state highways or other roads is high because the same road is shared by all, by both motorized vehicle users and pedestrians. Almost 73.5% of accidents has occurred on two wheelers, 13.1% has occurred on cars, 13.1% has occurred on jeeps and taxis. 6.9% has occurred on buses, 6.3% has occurred on auto rickshaws (Figure 12.2).

To our knowledge, over speeding is the primary cause of road accidents. There are some statistics for India which shows that in 2016, 73,896 people died because of over speeding. Due to rush, overtaking claimed 9562 people. There are many people who "drive under the influence of alcohol"; and this reason claimed 6131 people. Most people of India are aware of the correct reason of deaths, but there are 5705 deaths because of driving on the wrong side of the road. While crossing the road from the wrong way or ignoring the red signal, 4055 people died. The use of a mobile phone during driving took the life of 2138 people in India. Sickness or falling asleep or fatigue has claimed the life of 1796 people in India in 2016. Some other incorrect actions of drivers killed 17,943 people. Death because of "not wearing helmets" is 10,135. Four wheeler seat belt is a must. Five thousand six hundred thirty-eight people have died in a road accident in India just because of not wearing seat belts.

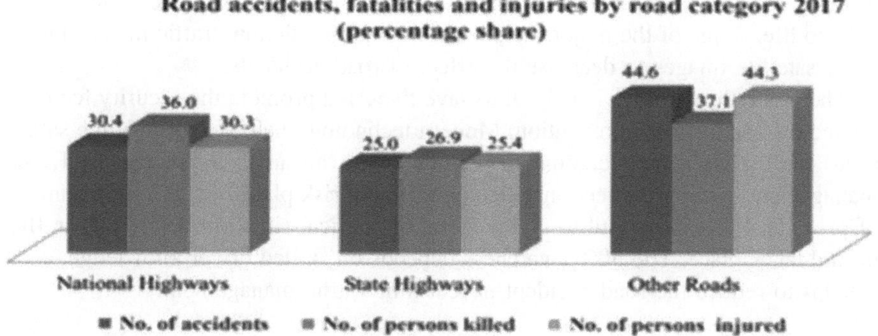

FIGURE 12.2 Total accidents on different categories of roads.

12.14 PREVENTION

Accidents on roads aggregate exponentially day by day. Hence, it is a huge concern for all the countries. According to research, almost 1.4 million people die each year due to accidents on road on a global scale. The type of reason for fatalities on roads can be controlled given it's not any disease. Only proper education and awareness is needed. There are some basic ways as stated below which will help to prevent this [8].

12.14.1 Using AI Techniques to Predict and Prevent Road Accidents

Current database system shows that more than 25,000 people are seriously affected due to rash driving and for not following the traffic rules. The use of phones, reckless driving, overspeeding on roads increase fatalities and injuries. Death is the main negative point which shows up as human error.

The possible secured process helps to decrease road accidents by using AI techniques. The number of accidents helps to account for the critical conditions based on the machine learning process. Most of the developing countries increase work efficiency to reduce the mortality rate due to road accidents. The proper data resources help to predict accidents before any untoward accident happens. The technical giants such as Tesla help to prevent the road accident with proper data analysis. The accidental cases indicate that self-driving is one of the major reasons for accidents in a congested area. The proper vehicle tracking system is used to identify the accidental vehicle. In UK, the government was able to change the traffic laws to secure the citizen lives. The autonomous self-driving has increased the accident rates on the road [9–11]. Artificial intelligence has helped to develop the infrastructure of traffic management and secure the lives of civilians.

12.14.2 Machine Learning Process Reduces the Life Risk

The assessment has indicated the machine learning process is effective to reduce the common road accident issues. The deep learning process and advanced machine learning models have on their databases more than 25 factors which can trigger risks to life. Some of the major factors are weather prediction, traffic management, using satellite images to decrease the effect of road accidents.

The prediction process is helpful to save lives and promote the security features related to road accident prevention. Modern technology helps to identify the safest route for the user to overcome the major traffic hazards. Given the vital risk management system has been installed on different risk platforms, it's the accuracy of prediction that is helpful to comprehend the critical situation before any traffic hazard takes place. The 50% awareness depends on skilled driver knowledge; and, it helps to reduce the road accident in real-time traffic management.

- **Engine maintenance** – Some accidents also happen from machine error. Bikes and cars are machines and sometimes, the vehicles as machines can lose control. The periodical maintenance of a vehicle will help us to avoid that [12,13].

- **Stop tailgating** – Driving behind trucks or other cars, people maintain little gap. In such cases, if the truck presses sudden breaks, then a collision cannot be avoided. Hence, a minimum gap is necessary according to the speed of vehicles. There must be a gap in which the speed can be controlled.
- **Be courteous** – The road does not belong to anyone; it is for everyone. The rule is also equal for everyone. Hence, the behavior with the other drivers must be polite [14,15] to prevent frayed tempers and consequent road accidents.

12.14.3 Avoid the Rush and Drunk Driving

As discussed, over speeding and drunk driving are some of the main causes of road accidents. If these two things can be avoided, then the number of accidents all over the world will be reduced for sure [16,17].

These are some causes of road accidents. According to the research, all of these are man-made. Hence, the solution of these catastrophies is also in the hands of humans.

12.15 LIMITATION

In every system, there are a few suitable information and a few limitations in it, which is obvious. In this topic "Road accident prediction using data mining" has some limitations. Some restrictions in technological work are obvious in many fields [18]. Most of these cases are monitored in Asiatic countries and also all over the world. The population of Asiatic countries is huge related to other countries, and this is the most important drawback of data mining in a road accident. Given this overpopulation, the data and related cases are much more voluminous, and the arrangement of it is very much difficult. Sometimes, the system might fail to maintain the data. Again, the illegal vehicles on the road are the main reason behind the limited influence of data mining. From the study, illegal vehicles as the cause of many accidents is the conclusion, but this drawback is not as important in other countries as in Asiatic countries [19]. The population was not counted in India after 2011, so it cannot be said what the population effects on road accidents are directly, but from an overall view and from the view of the states, the comparison between Asiatic countries and the world, the road accidents determine the effect of population. There is a gap in this study which can be taken up for further research. The number of illegal vehicles is not mentioned in any fields, and this study cannot mention the approximate value of the illegal vehicles in Asia or in the world, and the data on the states given are all outdated as mentioned above, so there is some further research that can be done in this respect.

The study-related limitation is that the data and information are very outdated, the graph and the statistical information are from before 4 or 5 years, recent incident and recent states data are the main gaps in this assignment [20]. Again, the suitable data is not available in the internet and, for this, there is some gap as the analysis which mainly should have stood on appropriate data and information is lacking for want of it, and it is considered as a drawback in the article. Because of the outdated data, the exact prediction is lacking. The appropriate amount of data on road accidents in recent years still unrecognizable, which the study didn't aim in this article, and because of using secondary data for the facts and research, there is considerable gap between observation and reality. This discussion and the sources of this work are based on secondary data and the study does not know whether this data is reliable or not [21]. The source of the data is a third party, and there is always a chance that there is erroneous data published, and for which the whole data analysis might be disproved. For there being a strong probability for this, there is gap for further research here. This article and the research is appropriate for the present time, and it will be outdated with time, because the facts will change over time; so, there is opportunity for further study and research. Again, we cannot rule out bias because of the secondary data. Given there is some probability of biases in the analysis and the data, and it is not obvious in secondary data; still it is a reason for further study to be done. Another limitation in this article is that there is a selected, small amount of data on road accidents and data on the states used. If this article's word count will allow to compare and analyze the data on a global scale, then there was possibility of a more appropriate prediction, which will show the actual picture of a road accident. The most common issue is the size of the sample in this work is very small, and from this lacuna, the research could not be done properly and the article cannot give sufficient weight with good data for analysis to take place, and there is a lack of clear point of view for this reason. For the outdated data and the limitation of sources, there is some gap that future work can take up. In this section, the study's limitations are discussed and once these gaps are fulfilled by future work, this article will provide a clearer picture for the prediction of road accidents in recent years [22].

12.16 RECOMMENDATION

Necessary steps must be taken to stop these mishaps. Because people are getting careless on roads, it's imperative to make them aware, for which education of civilians on strict laws are needed. But it also obvious that, in spite of having strict laws, people are not careful about maintaining laws. It shows a lack of implementation. There are plans and strategy, but the approach is not correct. To prevent road accidents, the first thing a country's government can do is to educate the countrymen. They should be made aware and should know that it is in our hands to change the turn of events to their benefit. There are not many things that people can control. But road accidents from human errors can be controlled. People always want to control those things which are uncontrollable. But they do not care about those things which are easy to control. A road accident is one of those things. If people are not aware, then there is no solution for it [23,24]. They cannot be controlled by any rule of law. It is up to them. The government or the other social work organizations can extend their helping hand by providing awareness. Only government can enforce awareness on laws and road accidents. If a single people raises his voice about it, then there will be more people with him. That is how the problem can be solved. The machinery error, which cannot be controlled all the time, is the only other thing that must be taken care of by people over time. Hence, there should be precautions for machinery error. A few steps can be taken, like periodical checking of cars and bikes. There should be organizations which can influence normal people to do so [25,26].

12.17 SIGNIFICANCE OF THE STUDY

The significance of this assessment is measured in critical situations. A government can take possible remedies to protect civilians' lives from road accidents. The study is helpful to gather practical knowledge about road accidents. A civilian knows how to handle any critical situations on the road [27]. The research has helped to analyze the basic skills and application to collect proper data from related case studies. The research has helped to build personal strength and skill ability from the relevant case study. The real issue has properly been described with different types of models, and possible risk factors which are mainly responsible for fatal injuries during road accidents are listed. The article also defines limitations which are lack of appropriate database management procedure impeding data mining due to the unavailability of the related case study. The article has mentioned the different kinds of risk factors which can easily affect the research study. The data procedure based on realistic data services for accurate data mining is needed. The data mining is helpful to predict the possible risk factors. The government can provide proper feedback to deflect the upcoming conditions. The government can measure the side effects of current status of road accidents. The traffic development authority collected the road accident data and secured it in the cloud file. The secured data is annually measured by the authorized authority to ascertain the possible risks [28]. The process is helpful in developed countries. It is possible due to knowledge and proper skill management to utilize critical conditions. The summary of the assignment has developed the basic skills of the students. The theoretical knowledge applies to develop critical conditions. The different models are applied to utilize the critical conditions and measure the possible risk factors which may occur during any kind of road accident. The provided methods help to improve traffic management and road safety to provide better services to the general public. Every government provides a huge amount to maintain public security.

The above case study has indicated one thing that the developing countries are facing critical problems due to overpopulation and lack of knowledge. Many of the private sectors have provided license without any kind of examinations. The personal ability not is tested properly. The many road accidents have occurred due to breakless and rash driving on the road [29]. The proper graphical representation highlighted the current condition. The developed country must be supported by third developing countries to provide modern technologies to improve the traffic infrastructure. The proper data collection and compilation procedure is most important to evaluate any base. The related model and applicable modern techniques can reduce the life-threatening road accidents. The modern data techniques can record the personal behaviors of the drivers on the road. The proper infrastructure is also needed to develop the road conditions, and the environmental factors may vary based on regional sectors. The proper decisions can help to maintain the traffic system; conditions of road depend on the proper material and technology provided for the security management procedure. The suitable data collection procedure helps to build proper data mining to ensure safety and security to prevent road accidents [6,7].

12.18 CONCLUSION

The country which has second largest road network in our world is India. Having the second largest road network, she is also in road accident proneness the most ahead. In well-developed countries, they use new technologies in their road constructions. While comparing with the city, the rural area displays high mortality in India. Statistical reports say factors behind road accidents are type and age of vehicle, age and gender of a person, type of road and zones, climatic conditions, etc. Everything is represented in graphical form for the better understanding of public to reduce the morality. Currently manual system is used by government; they are using data in ledger to analyze and to prevent the accidents. This report uses previous road accident records to predict the accidents which are happening frequently. Relations between the various road accidents are discovered by Apriori algorithm. The prevention of road accidents is done with the help of Naïve Bayes algorithm, Apriori algorithm from Data mining. This will help to avoid road accidents in future.

REFERENCES

[1]. Griffin, T. G., Young, M. S., & Stanton, N. A. (2017). Human Factors Models for Aviation Accident Analysis and Prevention. United States: CRC Press. Retrieved from: https://content.taylorfrancis.com/books/download?dac=C2015-0-81916-1&isbn=9781317120100&format=googlePreviewPdf [Retrieved on 21.10.2019]

[2]. Kjellen, U., & Albrechtsen, E. (2017). Prevention of Accidents and Unwanted Occurrences: Theory, Methods, and Tools in Safety Management. United States: CRC Press. Retrieved from: https://www.taylorfrancis.com/books/9781315120973 [Retrieved on 21.10.2019]

[3]. Dua, S., & Du, X. (2016). Data Mining and Machine Learning in Cybersecurity. United States: Auerbach Publications. Retrieved from: https://content.taylorfrancis.com/books/download?dac=C2009-0-26064-1&isbn=9781439839430&format=googlePreviewPdf [Retrieved on 21.10.2019]

[4]. Mastibekov, O. (2014). Leadership and Authority in Central Asia: The Ismaili Community in Tajikistan. United Kingdom: Routledge. Retrieved from: https://content.taylorfrancis.com/books/download?dac=C2012-0-13283-5&isbn=9781135006839&format=googlePreviewPdf [Retrieved on 21.10.2019]

[5]. Janani, G., & Devi, N. R. (2016). Road traffic accidents analysis using data mining techniques. Journal of Information Technology and Applications, 14(2). Retrieved from: http://doisrpska.nub.rs/index.php/jita/article/view/4346/4166 [Retrieved on 21.10.2019]

[6]. Madhumathy, P. et al. Smart transport system. International Journal of Industrial Electronics and Electrical Engineering, 5(1), 2017.

[7]. Sarada et al. Smart ambulance service system. International Journal of Applied Engineering Research, 10(55), 2015.

[8]. Gupta, M., Solanki, V. K., Singh, V. K., & García-Díaz, V. (2018). Data mining approach of accident occurrences identification with effective methodology and implementation. International Journal of Electrical and Computer Engineering, 8(5), 4033. Retrieved from: https://search.proquest.com/openview/feb083f841d05dd2258c4f3836ce8962/1?pq-origsite=gscholar&cbl=1686344 [Retrieved on 21.10.2019]

[9]. Sikdar, P., Rabbani, A., Dhapekar, N. K., & Bhatt, D. G. (2017). Hypothesis testing of road traffic accident data in India. International Journal of Civil Engineering and Technology, 8(6), 430–435. Retrieved from: https://www.researchgate.net/profile/Nk_Dhapekar/publication/326847372_Hypothesis_Testing_of_Road_Traffic_Accident_Data_in_India/links/5b69297a299bf14c6d94fb47/Hypothesis-Testing-of-Road-Traffic-Accident-Data-in-India.pdf [Retrieved on 21.10.2019]

[10]. Singh, M., & Kaur, A. (2016). A review on road accident in traffic system using data mining techniques. International Journal of Science and Research, 5(1). Retrieved from: https://pdfs.semanticscholar.org/8691/1421ecf8496a5f19b9ffd14ab2c15a459953.pdf [Retrieved on 21.10.2019]

[11]. Solanke, N. A., & Gotmare, A. D. (2018). Analysis of roadway traffic using data mining techniques for providing safety measures to avoid fatal accidents. International Journal on Future Revolution in Computer Science and Communication Engineering, 4(6), 45–50. Retrieved from: https://pdfs.semanticscholar.org/78e4/5549b0d5b383b228d8b2c736565ba305759b.pdf [Retrieved on 21.10.2019]

[12]. Muhammad, L. J., Sani, S., Yakubu, A., Yusuf, M. M., Elrufai, T. A., Mohammed, I. A., & Nuhu, A. M. (2017). Using decision tree data mining algorithm to predict causes of road traffic accidents, its prone locations and time along Kano–Wudil highway. International Journal of Database Theory and Application, 10, 197–206. Retrieved from: http://article.nadiapub.com/IJDTA/vol10_no1/18.pdf [Retrieved on 21.10.2019]

[13]. Nawrin, S., Rahman, M. R., & Akhter, S. (2017). Exploring k-means with internal validity indexes for data clustering in traffic management system. International Journal of Advanced Computer Science and Applications, 8(3), 264–268. Retrieved from: https://www.researchgate.net/profile/Shamim_Akhter5/publication/315966848_Exploreing_K-Means_with_Internal_Validity_Indexes_for_Data_Clustering_in_Traffic_Management_System/links/594829daa6fdcc70635a1bbb/Exploreing-K-Means-with-Internal-Validity-Indexes-for-Data-Clustering-in-Traffic-Management-System.pdf [Retrieved on 21.10.2019]

[14]. Kumar, S., & Toshniwal, D. (2017). Severity analysis of powered two wheeler traffic accidents in Uttarakhand, India. European Transport Research Review, 9(2), 24. Retrieved from: https://link.springer.com/content/pdf/10.1007%2Fs12544-017-0242-z.pdf [Retrieved on 21.10.2019]

[15]. Latha, G. S., VeereshBabu, D. V., & Thejraj, H. K. (2017). Prevalence of road traffic accident in children: Retrospective study in tertiary centre. International Journal of Contemporary Pediatrics, 4(2), 477. Retrieved from: https://pdfs.semanticscholar.org/6c54/b5f99168d2db67c1929d049a1f60e86e56f6.pdf [Retrieved on 21.10.2019]

[16]. Gupta, M., Solanki, V. K., & Singh, V. K. (2017). Analysis of datamining technique for traffic accident severity problem: A review. In Proceedings of the Second International Conference on Research in Intelligent and Computing in Engineering (Vol. 10, pp. 197–199). Retrieved from: https://www.researchgate.net/publication/317495696_Analysis_of_Datamining_Technique_for_Traffic_Accident_Severity_Problem_A_Review/link/5991d90baca27289539bafa9/download [Retrieved on 21.10.2019]

[17]. Hossain, M. S., & Faruque, M. O. (2019). Road traffic accident scenario, pattern and forecasting in Bangladesh. Journal of Data Analysis and Information Processing,

7(2), 29–45. Retrieved from: https://www.scirp.org/pdf/JDAIP_2019030416531017.pdf [Retrieved on 21.10.2019]

[18]. Shiau, Y. R., Tsai, C. H., Hung, Y. H., & Kuo, Y. T. (2015). The application of data mining technology to build a forecasting model for classification of road traffic accidents. Mathematical Problems in Engineering, 2015. Retrieved from: http://downloads.hindawi.com/journals/mpe/2015/170635.pdf [Retrieved on 21.10.2019]

[19]. Kaur, G., & Kaur, H. (2017). Black spot and accidental attributes identification on state highways and ordinary district roads using data mining techniques. International Journal of Advanced Research in Computer Science, 8(5). Retrieved from: https://pdfs.semanticscholar.org/f1c0/ed19e6a3087d48320a723b4d4d2757b4a4e0.pdf [Retrieved on 21.10.2019]

[20]. Satu, M. S., Akter, T., Arifen, M. S., & Mia, M. R. (2017). Predicting accidental locations of Dhaka-Aricha highway in Bangladesh using different data mining techniques. International Journal of Computer Applications, 165(12). Retrieved from: https://www.researchgate.net/profile/Md_Satu/publication/316999249_Predicting_Accidental_Locations_of_Dhaka-Aricha_Highway_in_Bangladesh_using_Different_Data_Mining_Techniques/links/591d3b0daca272d31bcb67ef/Predicting-Accidental-Locations-of-Dhaka-Aricha-Highway-in-Bangladesh-using-Different-Data-Mining-Techniques.pdf [Retrieved on 21.10.2019]

[21]. Souza, J. T. D., Francisco, A. C. D., Piekarski, C. M., & Prado, G. F. D. (2019). Data mining and machine learning to promote smart cities: A systematic review from 2000 to 2018. Sustainability, 11(4), 1077. Retrieved from: https://www.mdpi.com/2071-1050/11/4/1077/pdf [Retrieved on 21.10.2019]

[22]. Mazimpaka, J. D., & Timpf, S. (2016). Trajectory data mining: A review of methods and applications. Journal of Spatial Information Science, 2016(13), 61–99. Retrieved from: https://digitalcommons.library.umaine.edu/cgi/viewcontent.cgi?article=1084&context=josis [Retrieved on 21.10.2019]

[23]. Sriramoju, S. B. (2017). Review on Big Data and mining algorithm. International Journal for Research in Applied Science and Engineering Technology, 5, 1238–1243. Retrieved from: https://www.researchgate.net/profile/Shoban_Sriramoju/publication/321155155_Review_on_Big_Data_and_Mining_Algorithm/links/5a1140790f7e9bd1b2bf3e85/Review-on-Big-Data-and-Mining-Algorithm.pdf [Retrieved on 21.10.2019]

[24]. Yoo, J. Y., Lee, M. H., Aloyce, G., & Yang, D. M. (2016). Creating a Naïve Bayes document classification scheme using an Apriori algorithm. Advanced Science and Technology Letters (Current Research Trend of IT Convergence Technology IX), 133. Retrieved from: https://pdfs.semanticscholar.org/0a49/3dec160879033adcac19ccf9144a44eb19d1.pdf [Retrieved on 21.10.2019]

[25]. Le, C. C., Prasad, P. W. C., Alsadoon, A., Pham, L., & Elchouemi, A. (2019). Text classification: Naïve Bayes classifier with sentiment lexicon. International Journal of Computer Science, 46(2), 141–148. Retrieved from: https://researchoutput.csu.edu.au/ws/portalfiles/portal/30550232/30550129_Published_article.pdf [Retrieved on 21.10.2019]

[26]. Mahajan, N., & Kaur, B. P. (2016). Analysis of factors of road traffic accidents using enhanced decision tree algorithm. International Journal of Computer Applications, 135(6), 1–3. Retrieved from: https://pdfs.semanticscholar.org/9626/d71ec3187cc3c9ce10c091ce1ed7e5100272.pdf [Retrieved on 21.10.2019]

[27]. Pan, B., Demiryurek, U., & Shahabi, C. (2016). U.S. Patent No. 9,286,793. Washington, DC: U.S. Patent and Trademark Office. Retrieved from: https://patentimages.storage.googleapis.com/9f/58/db/8cff470bdf03ed/US9286793.pdf [Retrieved on 21.10.2019]

[28]. Krishnan, S., & Balasubramanian, T. (2016). Traffic flow optimization and vehicle safety in smart cities. Traffic, 5(5). Retrieved from: https://s3.amazonaws.com/academia.edu.documents/45875296/200_Traffic.pdf?response-content-disposition=inline%3B%20filename%3DTraffic_Flow_Optimization_and_Vehicle_Sa.pdf&X-Amz-Algorithm=AWS4-HMAC-SHA256&X-Amz-Credential=AKIAIWOWYYGZ2Y53UL3A%2F20191031%2Fus-east-1%2Fs3%2Faws4_request&X-Amz-Date=20191031T222913Z&X-Amz-Expires=3600&X-Amz-SignedHeaders=host&X-Amz-Signature=02ec1f3049d242b9a32c1c112cda58e7fe6e6ac17c4211c57a67d6aade4ee644 [Retrieved on 21.10.2019]

[29]. Atnafu, B., & Kaur, G. (2017). Analysis and predict the nature of road traffic accident using data mining techniques in Maharashtra, India. International Journal of Engineering Technology Science and Research, 4(10), 1153–1162. Retrieved from: http://ijetsr.com/images/short_pdf/1509888210_1153-1162-mccia960_ijetsr.pdf [Retrieved on 21.10.2019]

Index

H

I

J

K

L

M

N

P

R

S

T

U

V

W

Z